I0896471

MINIATURE PASSENGER HAULING RAILWAYS

MINIATURE PASSENGER HAULING RAILWAYS

PAUL CARPENTER

C Eng. BSc(Hons) FIMechE

PEN & SWORD TRANSPORT

AN IMPRINT OF PEN & SWORD BOOKS LTD.
YORKSHIRE – PHILADELPHIA

First published in Great Britain in 2025 by
Pen and Sword Transport
An imprint of
Pen & Sword Books Ltd.
Yorkshire - Philadelphia

Copyright © Paul Carpenter, 2025

ISBN 978 1 39908 387 4

The right of Paul Carpenter to be identified as author of this work has been asserted by him in accordance with the Copyright, Designs and Patents Act 1988.

A CIP catalogue record for this book is available from the British Library.

All rights reserved. No part of this book may be reproduced or transmitted in any form or by any means, electronic or mechanical including photocopying, recording or by any information storage and retrieval system, without permission from the Publisher in writing.

Typeset in Palatino 11/13 by SJmagic DESIGN SERVICES, India.

Printed and bound by Printworks Global Ltd, London/Hong Kong.

Pen & Sword Books Ltd. incorporates the imprints of Pen & Sword Books: After the Battle, Archaeology, Atlas, Aviation, Battleground, Discovery, Family History, History, Maritime, Military, Politics, Select, Transport, True Crime, Fiction, Frontline Books, Leo Cooper, Praetorian Press, Seaforth Publishing, Wharncliffe and White Owl.

For a complete list of Pen & Sword titles please contact

PEN & SWORD BOOKS LIMITED
George House, Beevor Street, Off Pontefract Road, Hoyle Mill, Barnsley,
South Yorkshire, England, S71 1HN.
E-mail: enquiries@pen-and-sword.co.uk
Website: www.pen-and-sword.co.uk

or

PEN AND SWORD BOOKS
1950 Lawrence Rd, Havertown, PA 19083, USA
E-mail: uspen-and-sword@casematepublishers.com
Website: www.penandswordbooks.com

FSC
www.fsc.org
FSC® C007031

CONTENTS

INTRODUCTION

This book is written by an enthusiastic model engineer who has spent a lifetime working through a wide range of practical and technical challenges to achieve the dream of building a miniature steam locomotive. These machines have always fascinated me. My parents were forever being dragged off to yet another miniature railway whenever I caught a glimpse of one.

Railways operate on technical principles which ensure that they function safely and reliably. In many cases these principles apply equally to both miniature and full-size practice, so a good understanding of the 'science' of railways is important, irrespective of size. The principles also ensure that the pleasures of operating a railway can be enjoyed with the least risk of injury or incident.

This book has three objectives:

1. To present practical information, technical and otherwise, necessary to build and operate a miniature railway by people of all backgrounds and skillsets.
2. To share the wonderful world of the miniature railways and the joy they bring to so many people, whatever age, background, or ability.
3. To encourage young people to get involved, learn new life skills, and discover the satisfaction of being 'hands on' and achieving things for themselves.

2½in Gauge The legendary *Tugboat Annie*, built by the Master himself, Lillian 'Curly' Lawrence - *LBSC*.

3½in Gauge. A beautiful GER Tank built by Gerald Sheed.

5in Gauge. BR Class 2 built by Richard Linkins and driven by son Alex.

No-one has ever built anything without making mistakes. Whilst inevitable, it is also very disheartening, so it is important to understand what went wrong, review the process, and try again. These hard learned lessons tend to be the ones that are not forgotten.

My dad was a toolmaker and had a friend who built miniature locomotives. We visited his workshop and I saw three 3½in gauge part-built locos under construction. For a ten-year-old kid, this was just magical, and I became determined to build my own engine one day. That chance came at my secondary school when a very enthusiastic metalwork teacher, Alan Sparks, gave me great support as I started building *Simplex*. With the help of Dad and Alan it first ran in 1972. *Torquay Manor* followed on and it won a Silver Medal at the 1997 ME show.

Recently, with a change of house, which my wife Clare selected because of the large rear garden, it became possible to realise a lifelong ambition to build a 5in gauge railway of my own.

This book describes the many and various challenges that building a railway presents. The book also considers several gauges as shown in the photos above and below.

7¼in Gauge. Chris Knibbs driving 4701 with a full train at Bentley Miniature Railway.

10¼in gauge. Original Basset Lowke *Royal Scot* on the Hastings Miniature Railway.

15in Gauge. Driver Alex Bishop in charge of *Samson* on the Romney Hythe and Dymchurch Railway.

Stephen Dowding and Keavy Mc Brown awaiting the road at the Bentley Miniature Railway.

Clare driving the local Passenger train with Speedy.

Chapter 1

SOME BASIC POINTS TO CONSIDER

1.1 Choosing a locomotive design to build

There are many published locomotive designs, so it is so important to have a thorough review of all that are available before committing to years of work. One of the best ways to do this is to join a club, talk to everyone and see for yourself how different designs perform on the track. Some designs are very old so their technical details may not represent current best practice, and others contain errors that have never been corrected. Talking to other engineers will quickly identify these.

Chapters 5 and 6 show how the valve gears of twelve locomotive designs can be 'optimised' using the latest techniques available.

Generally, it will take several years to build a locomotive, so the size and complexity of the design must be considered in terms of the following points:

1. Prototype or Freelance design? – will the engine please you when it's completed? – *Simplex* didn't please me!
2. Will it run on a raised or ground level track? – quite a difference in

Stuart Pierce driving the B4 at Jim's Railway.

Paul driving Peter Dickson's Saint at Bentley Miniature Railway.

Bob Driving his 28xx at Jim's Railway.

comfort for the driver, particularly for locos with long coal tenders. See photos showing a variety of driving positions.

3. How many passengers do you require the engine to pull? – private use or public passenger hauling?
4. Availability of drawings
5. Availability (and cost) of castings
6. Capacity of the home workshop
7. Capability, competence, and knowledge of the builder
8. Likely time available to build the engine.
9. Accessibility of the design for future maintenance
10. Overall weight for transporting to a track.

The issue of maintainability is often overlooked when assembling a locomotive. Unfortunately, this can become an issue later on and may involve significant dismantling to resolve otherwise straightforward issues. Engines with inside cylinders and motion can also be hard to lubricate and maintain. Large outside water tanks can also make access difficult.

Bob and Martin encouraging young drivers at Hall Place Bexley

Andy Blackwell driving his beautiful Deans Goods.

Peter Dickson taking on water for his beautiful GWR Saint Class.

1.2 Workshop Equipment

The heart of any workshop is the lathe. The popular Myford machine is ideal for a wide range of locomotive projects because of the range of accessories that are available for it. Three and four jaw chucks, together with a faceplate, will cover most of the basic turning operations. A set of collets would be excellent, and a vertical slide is a must in the absence of a milling machine.

Milling machines come in many shapes and sizes so the choice will depend on the size of components you expect to be milling. The following are key points that need to be considered before buying a Milling Machine:

1. The general robustness of the machine – many of the smaller milling machines lack rigidity and will chatter and vibrate when taking heavier cuts.
2. Check the vertical height available between the cutter and the machine vice. (Referred to as the 'daylight') Also check the longitudinal and lateral movement of the table. Limited travel in these three planes can cause much frustration and loss of time.
3. Check the size of morse taper through the vertical head and horizontal head, if fitted. This can limit the largest size of cutter that can be used.

A drilling machine is also essential in a workshop and again, versatility and capacity are keys point to look out for.

A bench grinding machine completes the essentials for machine tools.

The list of hand tools, taps etc is never-ending and can be acquired as needed and afforded.

1.3 Safety in the Workshop

Before suggesting some of the many basic skills that a model engineer requires, I must make a point about safety. Workshops can be very dangerous places and errors can easily lead to personal injury to fingers, eyes, and feet etc. In some extreme cases, I have known model engineers who have been killed by their own workshop machinery, so this must be taken very seriously.

1.4 Basic Essential Skills

The following list is not exhaustive but captures many of the key skills we need to build miniature locomotives:

- Reading a drawing
- Reading measuring instruments i.e. the Micrometer and Vernier.
- Becoming familiar with both imperial and metric systems of measurement
- Marking out
- Use of saws and files
- Sharpening drills
- Drilling and tapping
- Sharpening lathe tools of all the main forms.
- Basic turning operations.
- Basic milling operations.
- Fitting techniques.
- Heat treatment processes.
- Soft and silver soldering.

Chapter 2

GETTING STARTED

2.1 The Chassis

Irrespective of scale, it is important that the chassis acts as a reference point for the alignment of parts on the engine and therefore every effort should be made to ensure that the chassis is straight, level, and square. Always machine mainframes as a pair, along all edges. This allows measurements to be taken from any of these 'datum' edges and will be helpful as various components are fitted on.

Adding dowels allows the mainframes to be accurately aligned 'back-to-back' and are also useful for cylinder alignment if installed on the centre line of the motion.

Most of the chassis frame stretchers will not be visible on the finished model and so, wherever possible, take the opportunity to use heavier sections than scale. These allow the use of bigger and stronger fixing bolts, add strength, and provide much needed adhesive weight, whilst keeping the centre of gravity low.

Below left: Black 5 Chassis in 7¼ in gauge by Pat Harris.

Below right: Front view of the Black 5.

Cab view of the Black 5.

2.2 Horns and Axleboxes

The key for an engine having a long and trouble-free operating life lies in the alignment of the horns and axleboxes. Engines with poor alignment tend to wear out quickly as more clearance needs to be given in the bearings to accommodate the errors. Machining the horn apertures is best achieved by bolting the mainframes back-to-back and carrying out all machining in one set up. Swindon Works applied this method, machining up to five sets of mainframes in one setting using a large shaping machine. For miniature frames, bolting back-to-back is a very effective approach and the pair can then be machined to profile on a milling machine or equivalent.

Precision is so important when machining the horns and this is equally true when machining the axleboxes. The axleboxes should be a nice sliding fit in the horns. In full size, an axlebox should just slowly descend in the horns under its own weight.

Boring the axleboxes is achieved by use of a jig to ensure the required alignment, accuracy, and repeatability. Use reference edges to preserve a consistent offset from the rubbing faces of all axleboxes.

Miniature locomotive bearings are subject to many forces whilst working in a hostile environment. They invariably suffer wear and will eventually require replacing.

The roller bearing, axlebox and leaf spring suspension of the Black 5.

I have found 'Oilite' bushes ideal for main journals and other bearings. They retain oil, are readily available, not expensive and can be easily replaced.

2.3 Wheels, Axles and Crankpins

These three components require only straightforward turning skills, but it is important to plan the sequence of operations, completing as much machining in the one setting as possible. Then hold the work by the newly machined surfaces to complete other stages of the component.

Before starting machining on any casting, always check the sufficiency of metal to complete the later stages – many a good casting has been ruined by removing too much material in the initial set-ups, only to discover that there is insufficient metal to clean up the final stages.

When boring the wheels for the crankpin holes, always make a jig so that repeatability of the 'throw offset' is constant for all wheels.

Use Loctite 603 to retain the crankpins in the wheel bosses but turn half a dozen narrow grooves in the section being pressed into the wheel., to a depth of just 'a thou'. This ensures that the crankpin is a good mechanical and concentric fit within the wheel boss with the adhesive ensuring that the crankpin can withstand a high torque.

Axles should always be turned between centres in the lathe with a generous centre drill depth at each end.

Side view of the Black 5 Frames and running board.

Right: One of the Black 5 driving wheels.

Far right: Driving wheel crankpin showing the prototype return crank drive on the Black 5.

New Wheelsets for Bob's 28xx.

2.4 Quartering the Wheels

Before pressing the wheels finally on their axles, it is useful to make the coupling rods, as their centres can easily be determined by buttons – see Section 3.3. The button on the driving axle should be a good fit in the axlebox bore and locked in position. The remaining buttons must also be a good fit in the axlebox bore but require a clearance in the centre of the button. This allows the button to fit in their respective axleboxes and then locked in these positions. When the blank coupling rod is removed from the axleboxes, the buttons precisely locate where the rod must be bored. The buttons are then simply clocked up concentric on either the lathe vertical slide or mill and bored out.

The best way to set eccentrics is by using another jig. These control the valve events so accurate setting is very important here. The usual method is to calculate the horizontal offset from vertical of both the forward and reverse eccentrics. The jig then simply allows the eccentrics to role forward from vertical by this offset amount, ensuring correct valve timing.

Quartering wheelsets is best carried out between centres in the lathe. Two collars need to be turned that are a good fit on the crankpins and whose outside diameter is the same as the axle diameter which slightly protrudes from the wheel.

With the collars in place on the crankpins, and the wheelset between centres, put a square at the chuck end to set the crankpin in the vertical position, whilst the other end is set horizontally using packing and shims from the lathe cross-slide. The method gives excellent repeatability, which is critical.

2.5 Bringing the Parts Together!

Parts of the Black 5 awaiting assembly.

Newly painted Manor parts awaiting assembly by Roy Lipscombe of Midsomer Norton.

Roy's Manor coming together.

The following shots show Dave Owen's Foxcote Manor under construction:

Side view of 7822.

The 7822 Valve Gear viewed from below.

Front view of the 7822 bogie.

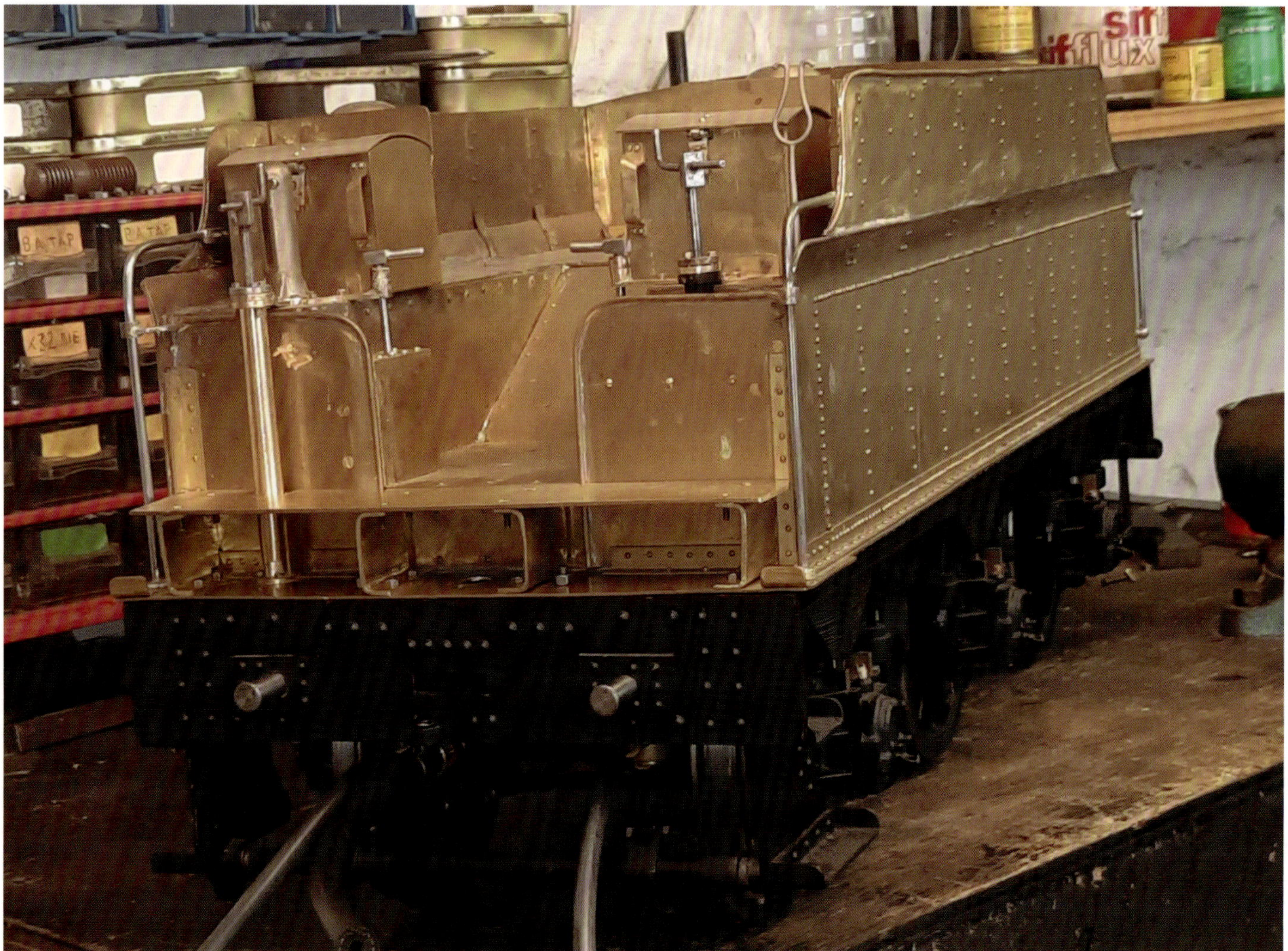

The completed 3500-gallon tender for 7822.

2.6 The Bogie

The following shots show progress on Pat's Black 5 bogie.

Side view of complete Black 5 bogie.

Assembled bogie for Pat's Black 5.

CHAPTER 3

CYLINDERS AND MOTION

3.1 Cylinders

The cylinders are the heart of a steam locomotive as they convert the thermal energy generated by the boiler into mechanical movement via the wheels. They are usually machined from castings, which can be made from either cast iron or gunmetal. My preference has always been for cast iron, but attention must be given to the bores to avoid them going rusty after use.

Steam entry to the main bores is controlled by valves and two types are used for miniature locomotives:

'Slide Valves' operate on a port face within a steam chest and require the machining of flat surfaces. There is more work in making this system than piston valves, but they are easier to make and will always be 'wearing in'. Steam, from within the steam chest, is admitted to the main bores by the valve exposing the port outside of the valve and is therefore referred to as 'Outside Admission'.

'Piston Valves' are the more efficient valves but require turning operations to fine tolerances. With use they do wear and cause leakage past the valves, but this can be managed using piston rings and sufficient lubrication. The ports admit steam from the inside faces of the valve and are referred to as 'Inside Admission'.

The bolting face should be used as a datum for all cylinder machining. The blocks can then be clamped either to a lathe cross slide for boring between centres or set up on a milling machine using a vertical boring head. A second machined datum should also be machined at one end at this stage.

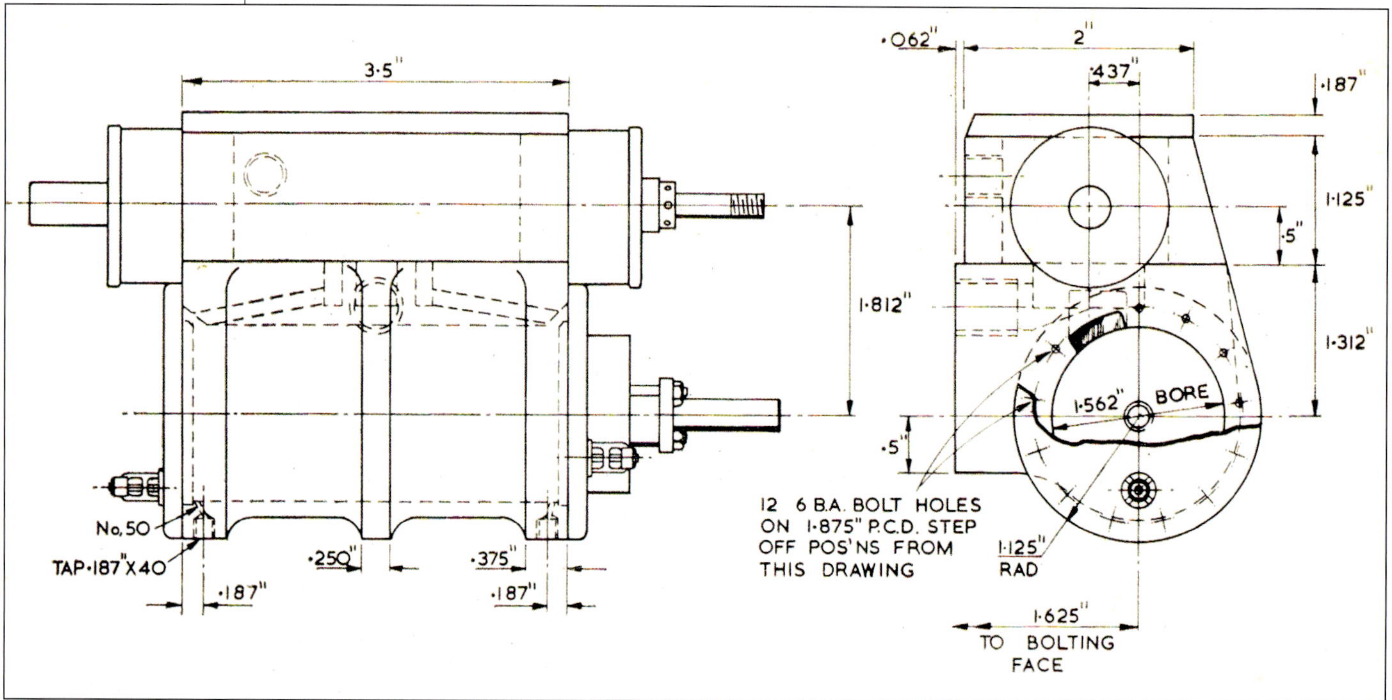

Manor slide valve cylinder.

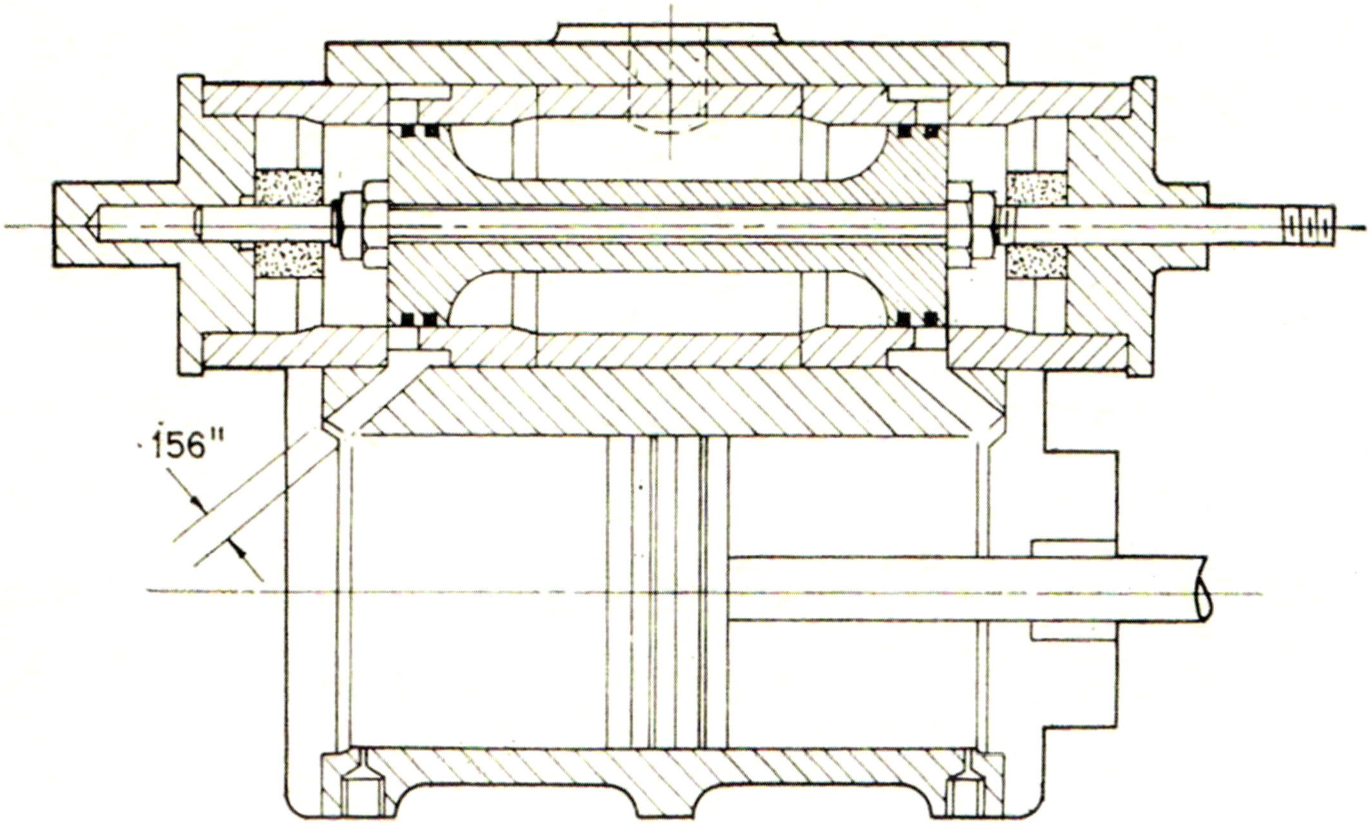

Manor Piston valve cylinder.

Clearly the bores must be parallel, and the surface finish good. Rigidity of the tool is essential to eliminate vibration and chatter and taking small cuts at a low speed helps. Honing bores provides an excellent surface finish giving years of trouble-free use.

Turning of the cylinder covers is a straightforward job but the slide bar supports on the rear covers require milling.

The main piston and the piston valve should be a good sliding fit in their respective bores with rings fitted where possible.

3.2 Slide Bars and Crossheads

The function of the crosshead and slide bars is to guide the piston in an axial plane whilst the connecting rod is exerting a bending force. The best supporting system for mounting the slide bars is to attach them between the rear cylinder cover and a motion bracket. Centrally supported slide bars can cause excessive wear in the cylinder bores.

It is important that the crosshead is a good sliding fit for the length of the slide bars. Once the slide bars are finally fixed and parallel, work can commence on the crossheads. Machine the piston rod spigot first and use this as the datum to set up the slide bars.

Crossheads can be machined from solid, fabricated or machined from castings, but in all cases, bronze slippers are essential on the sliding surfaces.

3.3 Coupling and Connecting Rods

The coupling and connecting rods are milled from rectangular bars of mild steel. Knuckle joints should be completed first, after which buttons

Bob's Jig for determining coupling rod centres.

can be loosely fitted on the rods and then inserted in the axlebox bores. The buttons now indicate the rod centres so should be locked in place, clocked up true on a mill, the buttons removed, and the rods bored for bushes. The connecting rods can be progressed in the same way.

I include a photo of a rather unusual looking jig by Bob Underwood which he uses to determine his coupling rod centres. It has the useful benefit of being adjustable and therefore able to be used for a range of engines and rod lengths.

3.4 Making the Motion

Generally, the various links of the motion are straight forward machining operations, but bronze bushes should be fitted wherever possible.

Expansion links present more of a challenge as these usually require a curved slot to be machined. A rotary table on the milling machine is the best way to machine these.

The following photos all feature cylinder and various motion parts for a beautifully built Black 5 in 7¼in gauge, made to works drawings by Patrick Harris:

Side view of a Black 5 cylinder and motion assembly.

Crosshead and coupling rod for the Black 5.

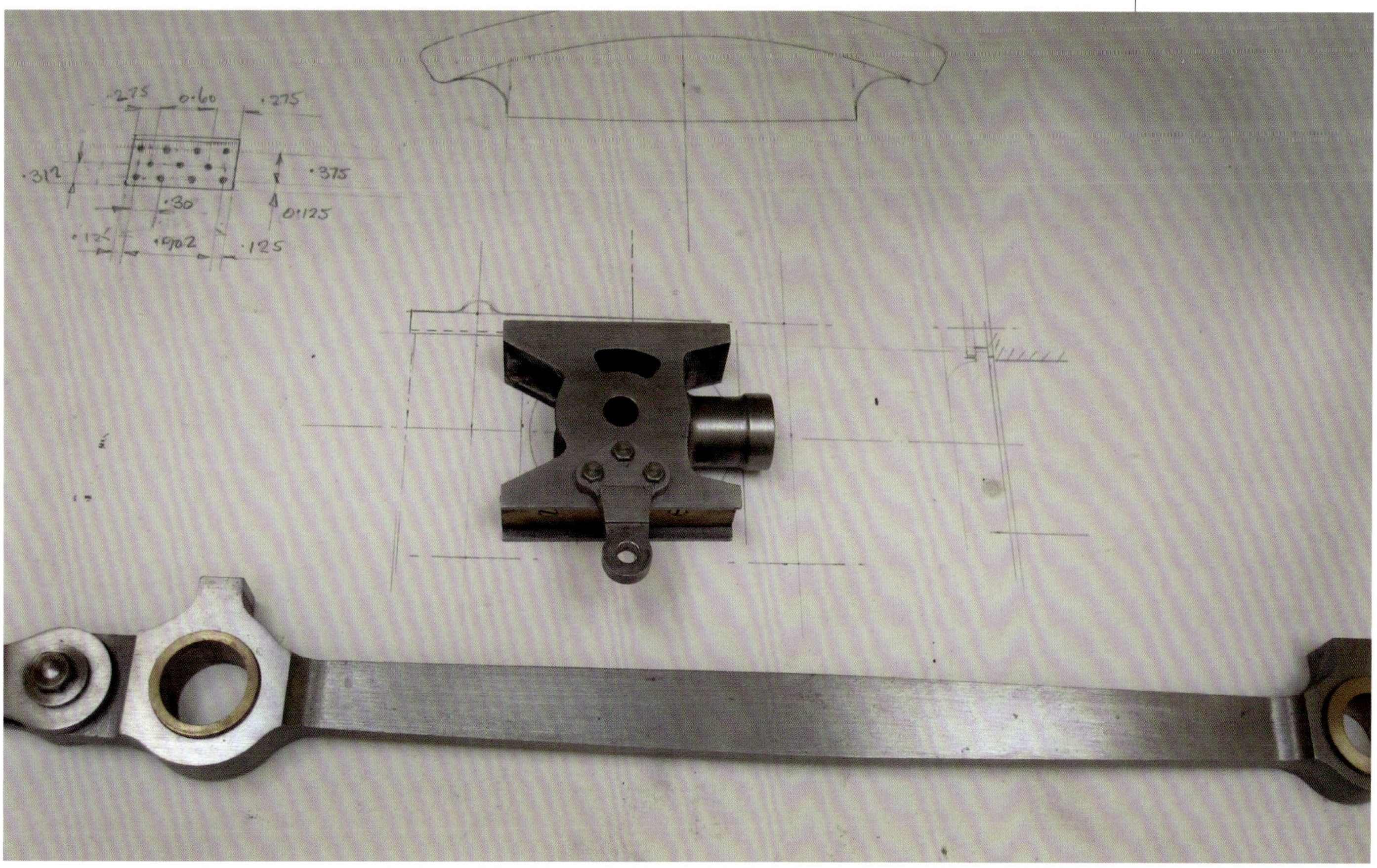

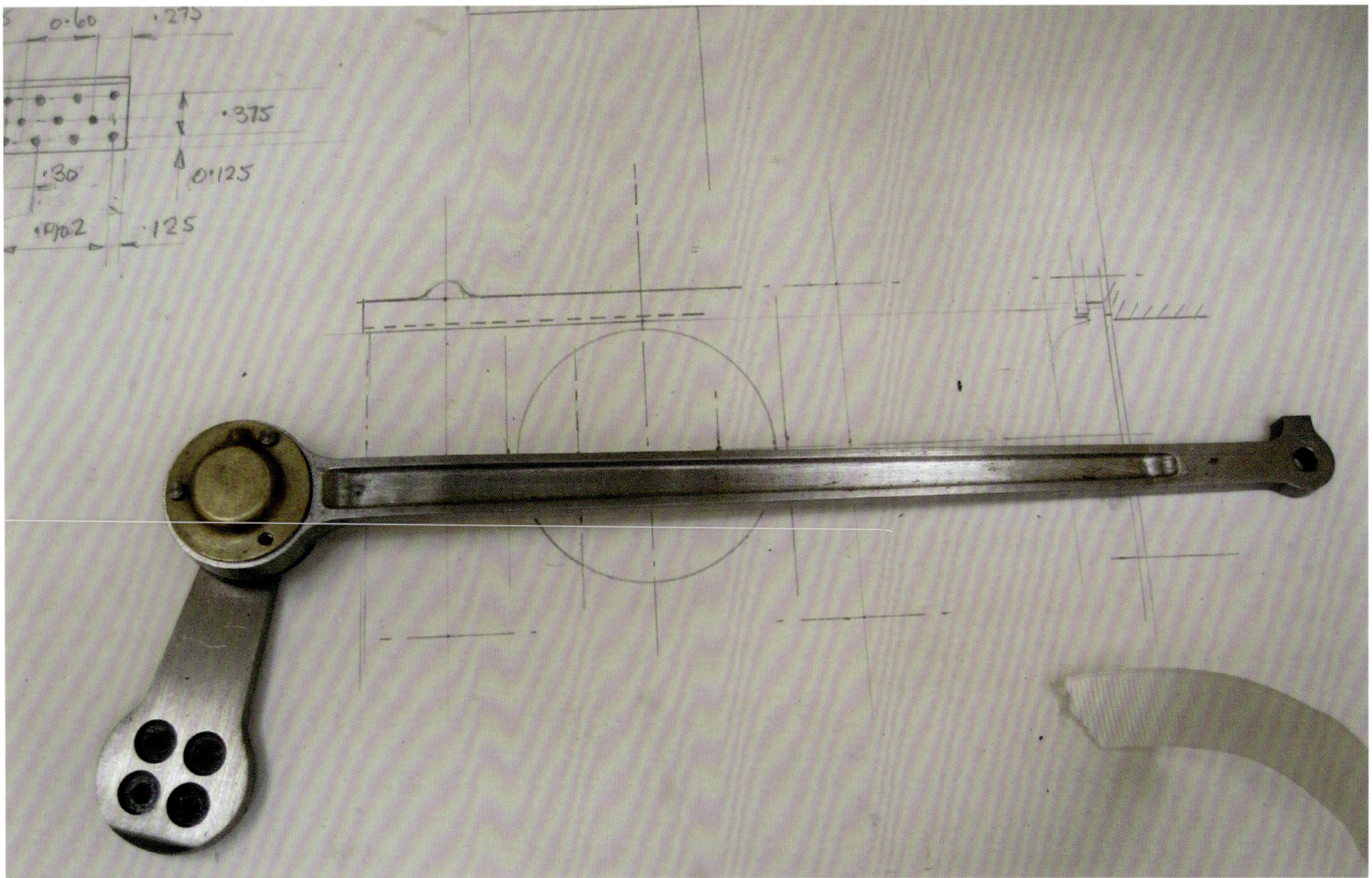

Return crank and eccentric rod of the Black 5.

Screw reverser for the Black 5.

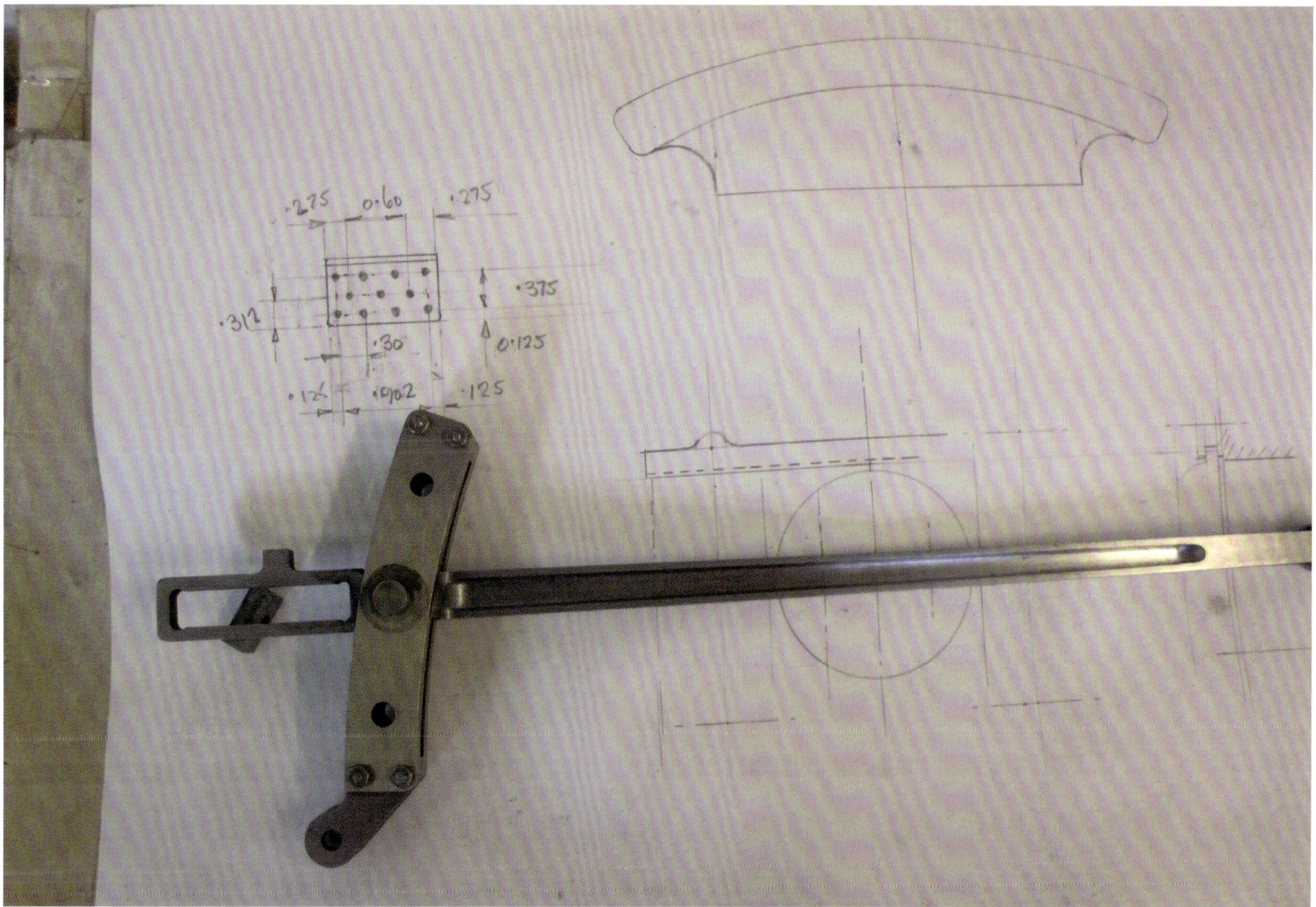

Expansion link and radius rod.

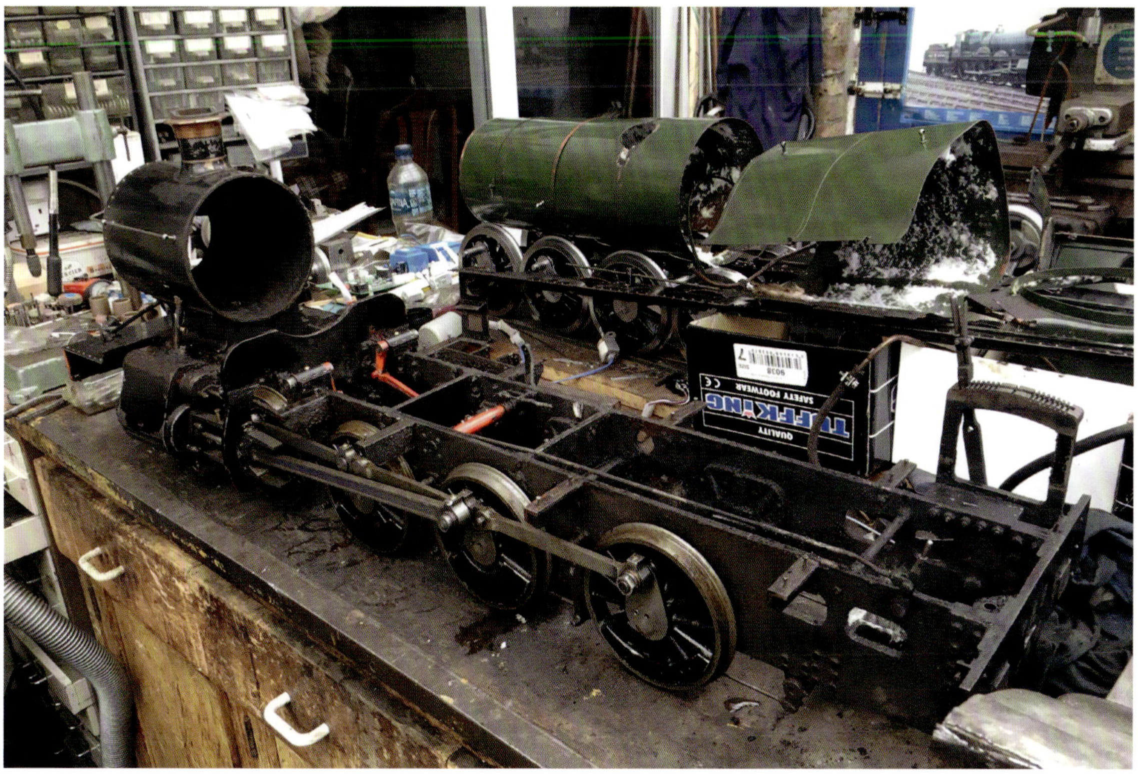

View of Bob's 28xx chassis having a 'Heavy General' overhaul.

A view of Bob's valve gear from his 28xx awaiting overhaul.

Chapter 4

DESIGN CRITERIA FOR STEAM LOCOMOTIVES

THE WORK OF JIM EWINS

Jim Ewins driving his loco 920 at a portable track run in the 1960s.

4.1 Background

Jim Ewins was an experimental and groundbreaking model engineer who lectured in Physics at Imperial College, London. He studied and experimented extensively, wrote many papers, and built locomotives to confirm his theories. He had great technical insights leading to defining principles that improved locomotive performance. I knew Jim for forty years and witnessed first-hand how his theories produced exceptional locomotives, the last of which I completed after his death.

This chapter touches upon several of the many design aspects that Jim shared through his writings in Club newsletters, the *Model Engineer* and *Engineering in Miniature* magazines. Jim's work first came to the attention of model engineers when his article on 'Testing a Locomotive Boiler' in *Model Engineer* March 1966 was published. Numerous technical articles followed but he always resisted writing his own book. This book records some of his work for model engineers of the future.

4.2 A Locomotive to Do What?

This simple question identifies the basic requirements for the performance of a design. If low speed, light load shunting is the task, then an 0-4-0 locomotive would be ideal. Full-size designers agreed and many industrial contractor's locomotives were built to this arrangement. 0-6-0 tank and tender engines were useful for local passenger and freight work, explaining why every region built so many.

Perhaps we expect all our engines to be 'high speed expresses' hauling heavy loads. These would require a larger wheel arrangement and diameter, with more adhesive weight. Full-size companies built many 4-6-0 type locomotives as well as 4-6-2s, 2-8-0s and 2-10-0s.

4.3 Balanced Design and Some Critical Relationships

Jim would commence his designs by defining the requirements. The following are some basic, but key points he considered:

- Coal performs in a miniature locomotive exactly as it does in full-size. It will have the same fire bed temperature and same air requirements. If air is fed at the same rate, it will burn coal at the same rate and reach the same temperature. This is the only common characteristic with full-size and therefore, the primary objective is to bring about this fire bed temperature and transfer the heat to create steam.
- The burning of the fire is controlled by the blast, and this will vary with the load, speed, track gradient and valve gear cut-off.
- There is an optimum for 'loading the grate'. This refers to how hard the fire is driven. Forcing the fire beyond a limit causes poor combustion, loss of

The plaque on the smokebox of Jim's 9F copyright Paul Parkinson ARPS.

efficiency, more rapid production of clinker and eventually blocked smoke tubes. Equally, not forcing the grate enough leads to a cool unresponsive fire and this fire will be inefficient, due to fuel not being completely burnt.

- The 'correct blast 'is that which meets the demands of the cylinders for the work to be done. The 'work done' is determined by the product of the engine's tractive effort and its speed. Both tractive effort and speed have limiting factors i.e., the adhesive weight and safety. It therefore follows that the adhesive weight of a miniature locomotive is the limiting factor for tractive effort and not the grate area. This is a significant difference from full-size practice. When Jim built his 9F (Lode Star) in 5in gauge, one of his objectives was to prove this point. That is why he made the mainframes from ½in plates and the engine carried a plaque on the smokebox stating this significant difference from full size.

4.4 'E' Numbers

Jim proposed a dimensionless constant he called the 'Engine Factor' which is defined as follows: -

$$\text{Engine Factor Ee} = \frac{\text{Swept Vol. of cylinders per rev.}}{\text{Grate Area} \times \text{Driving Wheel Dia.}}$$

He observed that locomotives that had the characteristics of a balanced and good performing engine, had Ee values within the range 0.12 to 0.25. Each of the variables in the formula can be changed to bring the Ee number within this range for any design.

Jim also observed that, due to 'scale factors', the miniature locomotive has a huge advantage over full size in respect of its heat producing potential, thus confirming that the power of a miniature locomotive is not limited by its grate area (as in full size) but its adhesive weight. He further suggested that an engine requires an optimum grate loading to utilise its full potential tractive effort. He identified this to be 6lbs of engine adhesive weight for every square inch of grate area.

Jim further observed that the 'scale effect' also applies to steam flow velocities i.e., steam velocities in a miniature engine are ¹⁄₁₂ of full-size, so for a scale length of a steam passage, the pressure drop along the passage is negligible. He went on to prove this with his design for 'Jimmy's Riddle' where all ports and passages were made deliberately small.

4.5 'E' Numbers – their equations and significance

Jim developed further relationships between locomotive parameters. He showed how the boiler proportions needed to meet the demands of the work done in the cylinders.

He called this the Boiler Factor Eb defined as:

$$\text{Boiler Factor Eb} = \frac{\text{Grate Area} \times \text{Tube length}}{\text{No. of Tubes} \times (\text{Tube ID})^2}$$

Finally, Jim defined the Engine Overall Factor Eo by combining Ee and Eb:

$$\text{Overall Factor Eo} = \text{Engine Factor} \times \text{Boiler Factor}$$

Much earlier, another experimental model engineer, C.M. Keiller, had identified an important relationship between the diameter of smoke tubes in a boiler and their length. He observed that the ideal relationship between these variables was given by the 'Tube Factor':

$$\text{Tube Factor} = \text{Kt} = \frac{\text{Tube length}}{(\text{Tube ID})^2}$$

Jim concluded that Eo is proportional to Steam Requirements divided by Capacity for Gas Flow

Summary of results for 'balanced engines'
Jim summarised his findings and proposed that a 'balanced and good performing engine' would have the following 'ideal' parameters:

Ee = 0.12 to 0.25 i.e. Average = 0.150
Eb = 80
Eo = 12
Kt = 80

Jim commented on the characteristics of an engine outside of the Eo range as follows:

If Eo was High, the steam demand was high compared to the available gas flow able to generate it. These engines would be more likely to be difficult to maintain steam pressure and more likely to run short of steam.

If Eo was low, there would be excessive gas flow, causing overdriving of the grate. These engines would be producing more steam than was needed, leading to early clinker formation.

The table below shows the analysis of the E Numbers for the twelve locomotive designs studied and developed in Chapter 6 by Simon Bowditch

4.5.1 Summary of E Number Results for twelve Engine Designs

Ewin's Numbers	Ee	Eb	Eo	Kt	Gas/Grate Ratio %	Adhesion Factor	Choke Diameter
Target Values	0.150	80	12	80			
Actual Values							
ENGINE							
Purley Grange	0.170	39	6.7	82	17	0.72	0.55
Princess Royal	0.089	83	7.4	103	10	0.38	0.81
Britannia Class 7	0.83	96	7.9	73	9	0.35	1.19
Lilla	0.377	63	23.8	76	18	1.60	0.79
Simplex	0.195	101	19.6	86	9	0.83	1.05
Speedy 15xx	0.191	46	8.8	82	20	0.81	1.04
Didcot 14xx	0.126	61	7.7	78	13	0.54	0.94
78xx Manor	0.172	70	12	79	16	0.73	1.10
BR Class 2	0.182	63	11.4	91	15	0.77	0.95
Dart 14xx	0.116	96	11.2	109	12	0.49	1.26
BR Class 2T	0.149	112	16.7	73	12	0.63	1.52
Black 5	0.129	136	17.5	81	13	0.55	1.75

4.6 Other Useful Formulae

Jim considered that for a balanced engine, the following range for the gas to grate ratio was very desirable:

$$\frac{\text{Gas Area}}{\text{Grate Area}} = 10\text{–}15\%$$

The full-size BR Standards had a very similar range of 14–16 %

Tractive Effort is given by:

$$TE = \frac{B^2 \times S \times P}{W}$$

Where B = cylinder bore in inches
S = cylinder stroke in inches
P = working pressure in lbs per square inch
W = wheel diameter in inches

4.7 Front End Draughting Design

Jim lectured in Vacuum Technology and brought great insights into the behaviour of gases in the smokebox and the requirements for efficient draughting. He realised that, due to scale velocities, it is not possible to replicate full-size practice. An example of this is the GWR 'jumper blast ring' which requires sonic velocity at the orifice which is unachievable in a miniature engine.

Therefore, the primary objective in a miniature locomotive is to create a sufficient degree of vacuum in the smokebox to suck the flue gases through the system, to match the combustion rate necessary to generate the required steam output. To achieve this, two conditions are required:

1. The ejector must be capable of handling the volume flow rate of the flue gas.

Full-Size draughting characteristics diagram by Sam Ell.

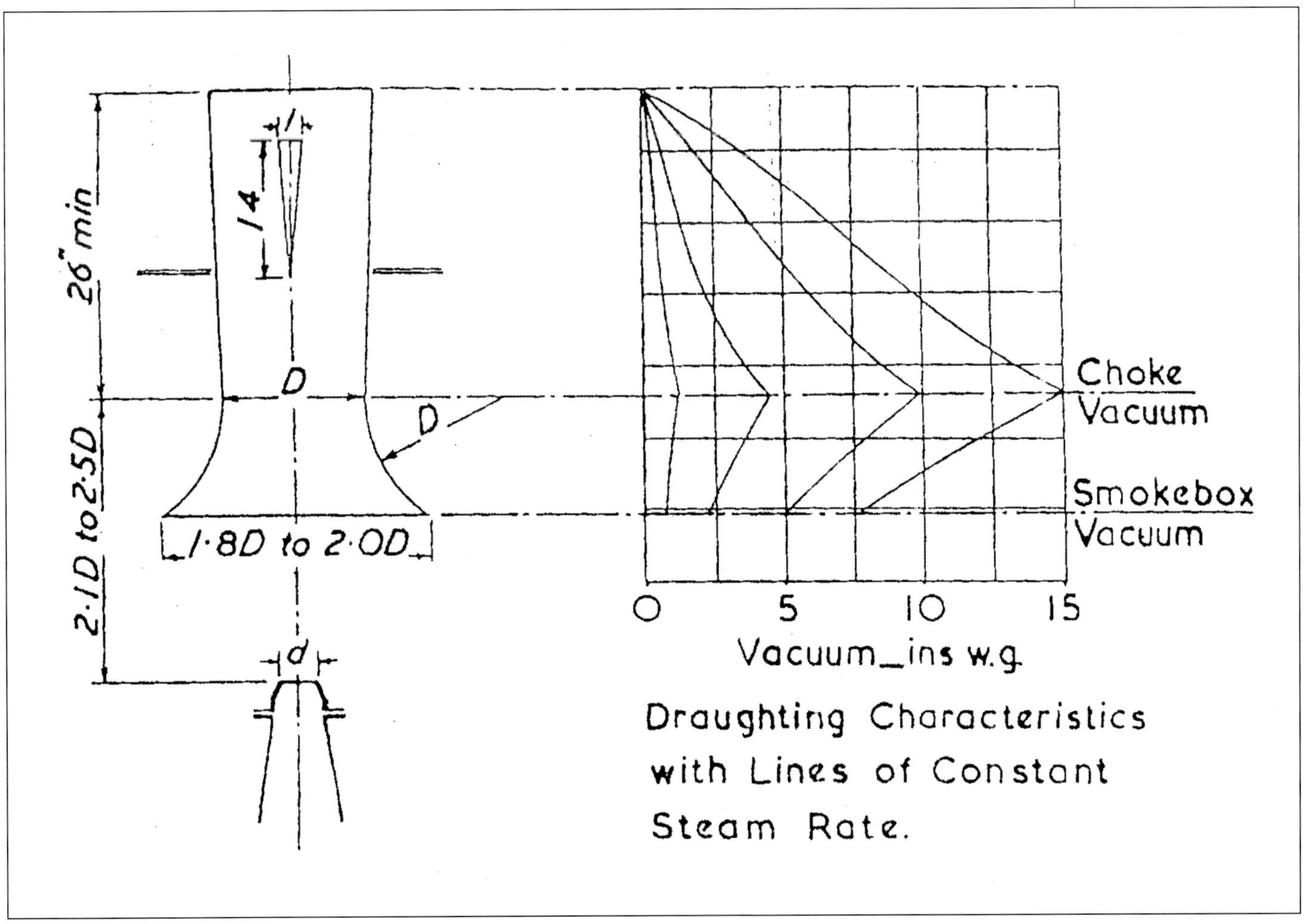

2. The ejector must cause sufficient depression in the smokebox to cause its volume flowrate.

The pumping effect of the ejector is derived from a surface effect between the cone of exhaust steam from the blast nozzle and the surrounding flue gas which gets entrained.

The area of the steam cone must therefore be related to the size of the fire producing the gases.

Size of the fire is related to the grate area and depth of fire.

The area of exhaust cone is related to the choke diameter and the position of the exhaust nozzle. The nozzle must not be too far from the choke otherwise the flue gas and exhaust steam will not be properly gathered.

The above allows the following relationships to be stated, based on ejector theory:

1. Choke tube length should be around 7x its diameter
2. The blast pipe nozzle should be 1.5 x the choke tube diameter below the point where the flare of the petticoat meets the choke tube.

As a rough guide, the choke area should be approx. 1⁄20 of the grate area i.e. choke tube diameter D = ¼ square root of the grate area. Full size Practice for the BR Standards used the following relationships: -

Throat area	=	3-4% of the grate area
Taper in the chimney	=	1in14
Adhesion factor	=	4-5

4.8 Superheating

The use for superheating in full-size steam locomotives is driven by the need for efficiency and keeping the amount of coal burned to a minimum. Steam tables clearly show that the energy in a unit volume of steam increases significantly with the increase in its temperature above that of its corresponding working pressure.

In miniature locomotives the need for superheat is equally important but for a different reason. Here cooling through the cylinders is disproportionately high due to the 'Square-Cube Law'. To explain this using a 5in gauge loco, our engine will be 1⁄12 linearly of the full-size, so an area in our model will be 1⁄12 squared = 1⁄144 of the prototype and its volume will be 1⁄12 cubed = 1⁄1728 of the prototype. For our miniature engine cylinders this means that the heat losses from a unit volume of steam enclosed by a cylinder wall area will be 12 times greater than the prototype. Not only is the heat loss disproportionately high, but there is also the potential risk for condensing some of the steam. The mainframes and other steel chassis parts further exacerbate this loss. Also, wet steam does not flow as freely as dry steam through the ports and passages. This helps to transfer heat to all the surfaces in contact with the wet steam hence the increase in cylinder cooling.

Steam can only be kept dry in the cylinders by starting with it at a higher temperature. The amount of water used to produce a given volume of steam also reduces with temperature, so a high superheat also requires less water to be boiled and less coal to be burned. These reductions allow the engine to steam for longer before issues with the fire clinkering occur.

There are two types of superheaters, the 'flue tube' type and the 'radiant' type. Full-size engines always use the flue tube type due to the high velocity of the exhaust gases as well as the steam inside the elements which creates a sufficiently high degree of superheat.

In miniature engines the flow of the exhaust gases is much lower due to the 'scale effect' and so the only way of ensuring superheated steam is to use the radiant type of element. Radiated heat is proportional to the difference in temperatures between the fire and the superheater element to the power of four. Clearly this will vary depending on how hard the fire is being driven but suggests that a very bright and hot fire is desirable.

The heat transferred to the steam from the superheater is described by the 'Reynolds Analogy' which states that the heat transferred to the steam through the wall of the superheater is directly related to the friction experienced by the steam in passing through the element. Jim used to flatten his superheater elements to increase friction and therefore the heat transferred.

4.9 Other Jim Ewins Engines

Jim built a very impressive number of locomotives and constructed them to highlight his scientific principles.

He first built 920. She was basically an LBSC Minx design which Jim extended to an 0-6-2 Tank. It was designed and built for hauling heavy passenger trains. Jim was a quick builder, and the number 920 represents the number of hours it took to construct.

His Schools Class loco was built to works drawings and is currently on permanent exhibition at the Milestones Museum on the outskirts of Basingstoke.

Jim driving *Lode Star* and hauling 89 passengers.

His 9F (*Lode Star*) was a magnificent engine which he featured in a series of articles in *Engineering in Miniature* – see Bibliography. He designed it to pull exceptional loads (89 passengers on one occasion).

Jimmy's Riddle was built to prove that small ports, valves, and passages were not detrimental to an engine's performance.

The Z Class was an engine that Jim just liked. He had seen them as a boy on his local railway and liked the design! He fitted an original valve gear to this engine, using a Scotch crank to drive the valves. His intention was to create pure simple harmonic motion of the valves, which he achieved.

H Class locomotive was his first 7¼in gauge engine.

A LSWR B4 Tank was his final engine which was only half finished when he died. I spent several years completing this engine in his memory.

The photos below show some of the locos Jim built, as well as visiting engines to his track.

Front-angled view of the B4 at the Midlands Show.

B4 on display at the Midlands Exhibition.

Jimmy's Riddle. (*Paul Parkinson ARPS*)

The plaque carried on the B4. (*Paul Parkinson ARPS*)

Jim's magnificent 9F.

Jim's Z Class running on his railway. (*Paul Parkinson ARPS*)

Busy day at Jim's track. Stuart is steaming the Z Class, Jack awaits the road and 7823 is on the mainline. (*Paul Parkinson ARPS*)

Paul driving the B4 at Jims. (*Paul Parkinson ARPS*)

Stuart entertaining the hosts at Jim's Railway.

John Bentley driving the B4 at Jim's Railway.

Chapter 5

AN INTRODUCTION TO THE COMPUTER SIMULATION OF LOCOMOTIVE VALVE GEARS AND VALVE EVENTS

BY SIMON BOWDITCH

Model Engineering has been blessed with several outstanding engineers over the years and I have been fortunate to work with several of them. I knew Don Ashton for almost twenty years, and he was undoubtedly an expert in many aspects of locomotive engineering, but especially valve gears. His knowledge and insights were truly inspiring. More recently, I was most fortunate to meet Simon Bowditch, who was not only a close friend of Don's, but also an expert on the 'black art' of valve gears. Who better to invite to share with us his great insights and outstanding, analysis and research, acquired after many years of study?

Simon Bowditch.

Over to Simon

Over the years much has been written about the design of model locomotive valve gears and the correct way to set the valves to achieve the performance sought by the designer. Some writers have resorted to more advanced levels of mathematics to demonstrate their theories, while others have oversimplified the process, leading model engineers up a blind alley.

Computer simulation allows model engineers to assess, test and adjust the parameters and performance of a valve gear before any metal is cut. The two packages referenced here have been developed over recent years and provide essential information to optimise and fine tune most valve gears.

When a newly built model locomotive fails to deliver the expected performance, model engineers frequently resort to various expedient measures to avoid tackling the valve gear issue which lies at the root of the problem. These measures often include better fuel, improved grates and ashpans, brick arches, elaborate superheater arrangements and modified exhaust systems. All these can make the boiler produce more steam, but until the valve gear design and setting are optimised to achieve accurate distribution to the cylinders, the additional steam production will be wasted. A lack of knowledge of how to achieve optimal valve events often deters builders from addressing these issues.

I have read many books about the development of the steam locomotive since its inception. Some railway companies were slow to adopt long lap/long travel valves and, in consequence, model

Don Ashton with Hairy Biker Dave Myers and Paul at Llangollen during the filming of 4709.

Don and Paul with Hairy Biker Si King at Llangollen for the 4709 documentary on the engine's construction.

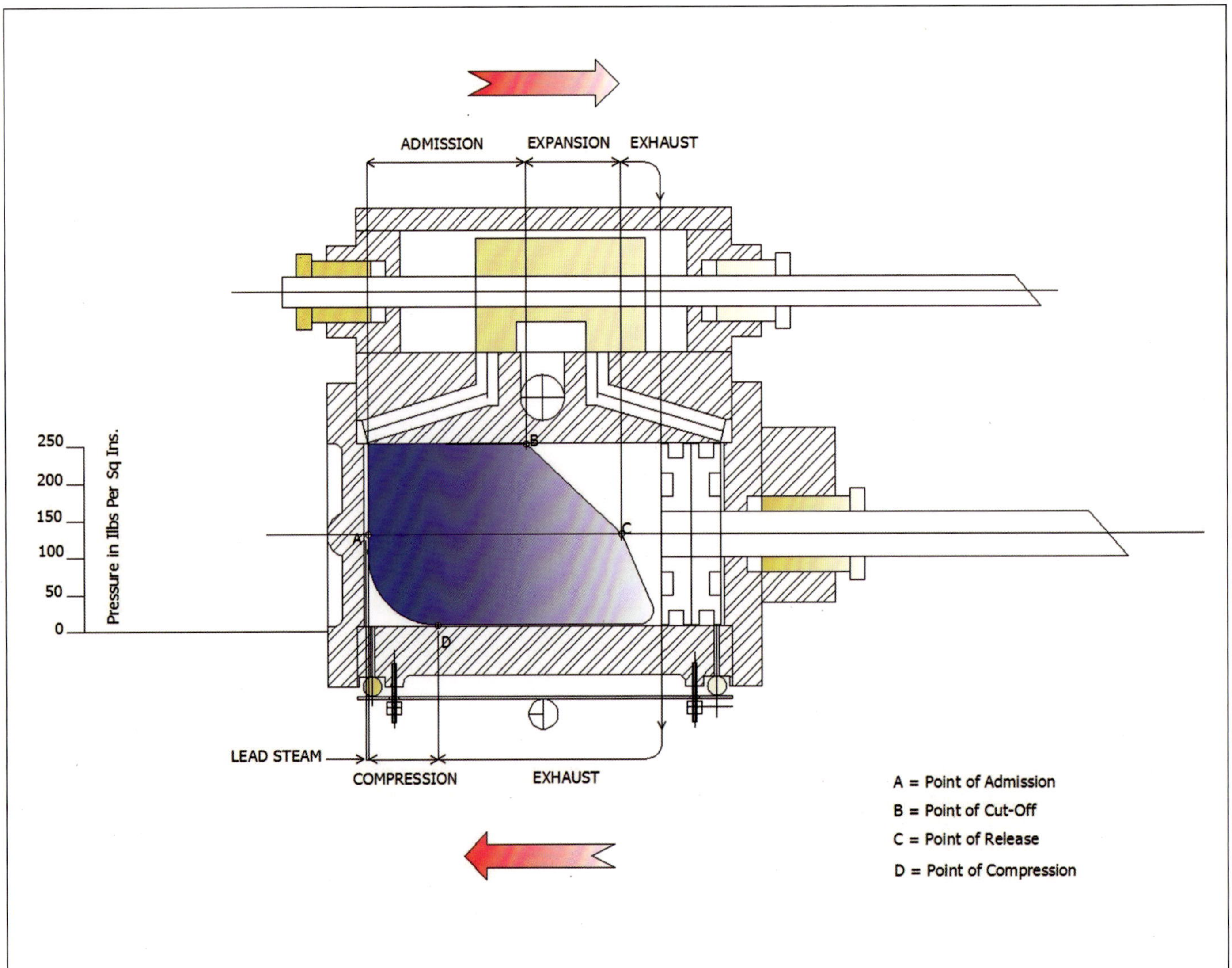

Diagram showing the distribution of steam on one side of the piston for a double stroke on a FULL-SIZE locomotive.

engineers became imbued with the idea that these proportions were the cornerstone of the miniature locomotive. Many contributors to model engineering journals encouraged us to believe that this was the way forward. However, I have come to realise that they were forgetting two important facts.

Firstly, when long valve laps and large ports are employed in conjunction with a full forward / reverse gear cut off as high as 85–90%, the valve events at shorter cut-off settings can become significantly distorted.

Secondly, the steam cycle in miniature locomotive cylinders is adversely affected by the low temperature of the steam produced by the model boiler. Despite attempts at superheating the steam before it enters the cylinders, most of the energy is lost as its heat is rapidly conducted away through the metal cylinder castings into the frames and surrounding atmosphere. The result is the moment after the valve closes the port at the point of cut-off, very little further work is done by the steam. In a full-size locomotive, with a steam temperature of 400ºC or more, the slowly diminishing steam pressure continues to propel the piston forward as the steam expands. In the case of a miniature locomotive with a working pressure of 80psi, the temperature of the steam could be as low as162ºC. Once the valve reaches the point of cut-off and steam ceases to be admitted to the cylinder, the temperature and pressure of the remaining steam plummets. By the time

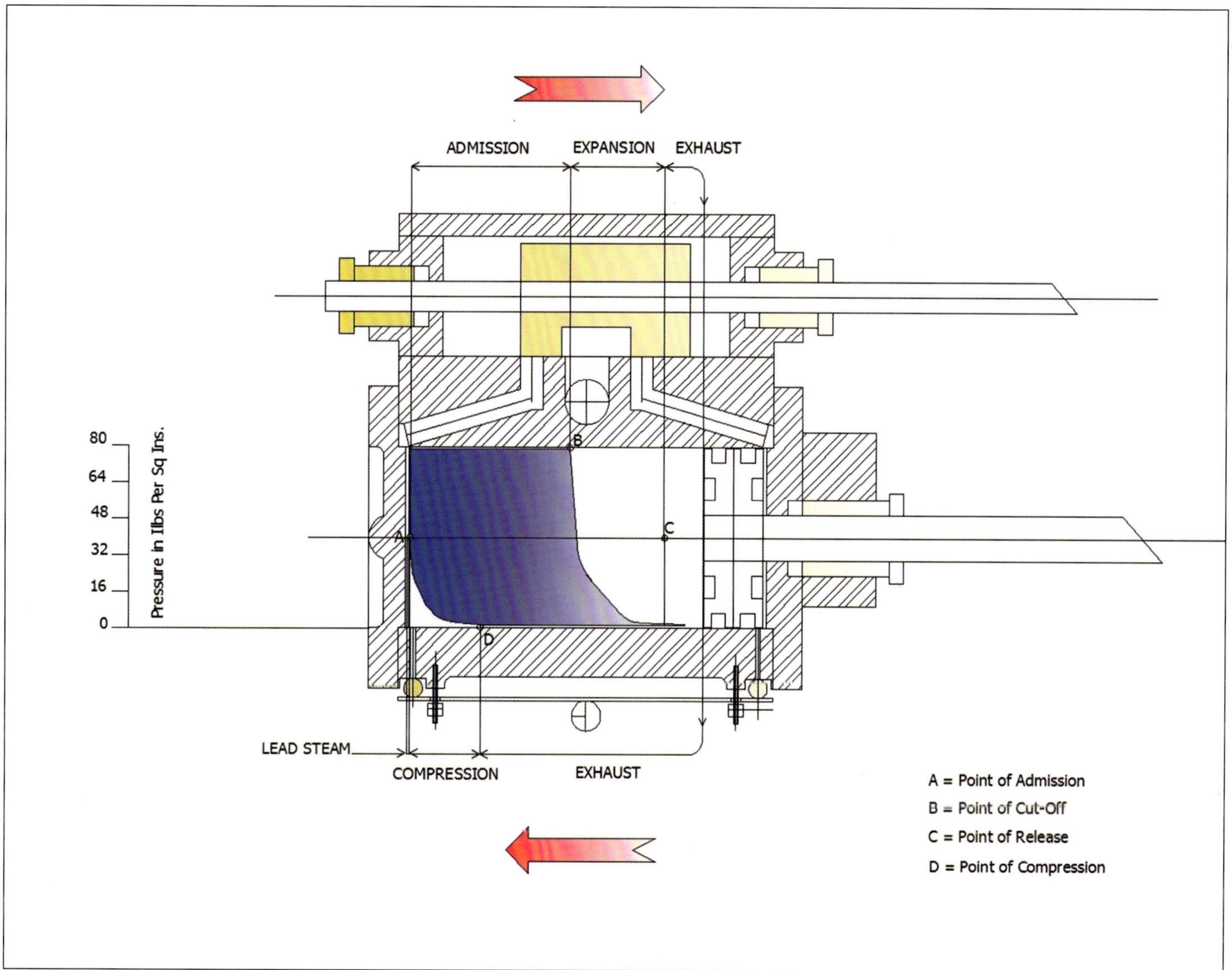

Diagram showing the distribution of steam on one side of the piston for a double stroke on a MINIATURE locomotive.

the point of exhaust is reached, little if any further work has been done by the steam. The whole expansion phase of the cycle never happened and the remaining heat and pressure in the cylinder contributed little to the task of powering the piston forward. Therefore, in the case of miniature cylinders, the action of the steam might be viewed as a 'pulse' of pressure on the piston rather than the sustained high pressure/high heat 'charge' found in a full-size locomotive.

Another phenomenon which significantly affects miniature locomotive cylinders is the condensing steam at the end of the stroke forming incompressible water droplets. Effective superheating can address this issue, but if these droplets are not exhausted from the cylinder, they could cause a hydraulic lock between the piston and cylinder cover on the return stroke.

In the early editions of his booklets published in the 1970s, Don Ashton clearly stated that the valve travel must be optimised to achieve symmetry of the valve events. It is only since computer valve gear simulation programs became available that Don's theories could be put into practice.

Don was one of the first to produce a valve gear design and simulation program using a Microsoft Excel spreadsheet. In the 1980s, Professor Bill Hall demonstrated his modified 5in gauge model of LBSC's design 'Speedy'. His 'Speedy' really outclassed all previous versions because the valve gear had been redesigned using his own

Simon driving Glyn Wells's 7¼in gauge *Sweet Pea* 0-4-0 Saddle Tank on the Squaredoch and Deskford Railroad in Moray. Simon's wife Loraine is enjoying the ride as passenger.

Simon driving Brian Remnant's IMLEC winning 5in gauge locomotive at Pembrey. Brian is riding as passenger.

valve gear simulation software. The revised design featured narrower ports and shorter valve laps creating greater valve event symmetry. Other innovators were also designing new simulation programmes e.g. Dr Allan Wallace in Australia and Charlie Dockstader in California. Both produced excellent software written for the DOS platform and I was delighted to explore these exciting programs on my own home computer. Unfortunately, many model engineers have been reluctant to learn and use these brilliant new tools.

My first practical application of this new technology was helping a friend redesign the engine for his steam launch. The valve gear was well worn and could not be relied on to allow reverse to be used for stopping. Consequently, new eccentrics and an expansion link were made to dimensions suggested by the program. When the launch was tested on water, its performance in both directions was a revelation to both of us.

Although Don Ashton, Professor Hall and Charlie Dockstader are no longer with us, their work remains a brilliant legacy for model engineers everywhere.

Valve gear simulation software is freely downloadable from the following website: http://www.donashton.co.uk/html/downloads.html. I would suggest a good starting point would be Professor Hall's programs for Stephenson and Walschaerts' valve gear. The data input is minimal, and the programs provide numerical and graphic output, together with the option of an animated graphic showing how the linkage and valves of your specification are working. The software is showing its age, but it gives a clear picture of what is happening.

Below is an example of how I tabulated the results for *Simplex* with the reversing arm angle of 8 degrees.

There are four columns giving the percentage values of four important stages of the steam cycle in the cylinders. You will also notice that each cell has been colour-coded green to indicate that the values for the front and back stroke of the piston of each stage of the cycle are within 2% of each other.

The examples in the next chapter and appendices contain varying values and the colour is helpful in identifying reverse lever settings where the values are lacking in symmetry.

The following colour-coding is used to highlight and score phases of the steam cycle where results are becoming asymmetric.

The results which fall within 2% of each other are coloured green and score 4 points. 2–3.5% are coloured yellow and score 3 points. Entries achieving 3.5–5% asymmetry are coloured amber and score 2 points. Anything over 5% is red, indicating that improvements are desirable.

The Valve Event Tables in Appendix 2 present the results for twelve locomotive designs and the scores have been totalled for each column and presented as percentages.

So, what can be considered a 'good result'? Previously, designers have concentrated on getting the full gear values correct in terms of equal cut-off percentage, equal port opening and equal

Key to Colour Code used to identify Asymmetry Issues found in Valve Event Charts	
0.0% to 2.0% Asymmetry	4
2.0% to 3.5% Asymmetry	3
3.5% to 5.0% Asymmetry	2
5% or more Asymmetry	1

Colour-Coding / Scoring System.

Prof. Bill Hall's Walschaerts Valve Gear Program. Valve Events for Simplex with Reversing Arm at 8 degrees														
Outside Admission Walschaert's Valve Gear fitted to 5" Gauge Martin Evans "SIMPLEX"0-6-0 Tank														
Notch	Angle	Travel	Compression		Admission		Cut-Off		Exhaust		Port Opening		Lead	
			Front	Back	Front	Back	Front	Back	Front	Back	Front	Back	Front	Back
5	8°	0.401"	14.72%	13.71%	0.206%	0.173%	57.69%	57.67%	86.29%	85.28%	0.072"	0.079"	0.012"	0.012"

lead. They assume that the valve gear will then link up, allowing the driver to control the engine on the reverser as well as the regulator. The results shown in Appendix 2 reveal the flaws in their assumption, but then provide the information necessary to make improvements in the locomotive's performance.

Prof. Bill Hall's Walschaerts Valve Gear Program. Valve Events for Simplex in the running position															
Outside Admission Walschaert's Valve Gear fitted to 5" Gauge Martin Evans "SIMPLEX"0-6-0 Tank															
Notch	Angle	Travel	Compression		Admission		Cut-Off		Exhaust		Port Opening		Lead	
			Front	Back	Front	Back	Front	Back	Front	Back	Front	Back	Front	Back
3	3.4°	0.300"	29.48%	29.02	0.783%	0.673%	23.87%	25.44%	70.98%	70.52%	0.024"	0.026"	0.012"	0.012"

Tarn Beck - another of Gerry Clark's great locos. He is seen coupling up at the 2014 7¼in Gauge Society AGM held at the East Somerset Society of Model and Experimental Engineers club track at The Royal Bath and West Showground near Shepton Mallet.

Simon stops *Trad Mawr Paddy* while the owner and builder, Glyn Wells, tops up the tender with water. Image taken on Glyn Wells' Squaredoch and Deskford Railway in Moray, Scotland.

In my experience, it is better to set the valves in a running position closer to mid gear, so that you can enjoy running the locomotive at the shortest cut-off that is appropriate for the load being hauled and the gradients on the track.

As shown in the table on page 54, I would suggest that with the reversing lever set to a cut-off of 25%, the values in the four columns would look like this:

At 25% cut-off, the front and back values should be within 2% of each other.

At 25% cut-off, compression values should be between 28.5% and 31.5%.

Admission values should be between 1% and 1.5%.

Exhaust values should be between 68% and 71%.

You might question the small amount of port opening recorded at this setting, but I can assure you that many efficient 5in gauge locomotives perform very well with a 0.025in port opening at 25% cut-off.

Also note the value of the Lead which is 0.012in – the thickness of a tram ticket, in LBSC's parlance!

CHAPTER 6

ANALYSIS BY SIMULATION OF TWELVE PUBLISHED LOCOMOTIVE DESIGNS

ALL MODIFIED DESIGNS ARE BY SIMON BOWDITCH, UNLESS OTHERWISE STATED

6.1 2.5in Gauge Locomotive Valve Gear Analysis

6.1.1 GWR 4-6-0 Purley Grange by LBSC

Improvements to the outside admission Stephenson's Valve Gear by Don Ashton

The 'Grange' class emerged from Swindon in 1936 and LBSC commenced his series in the *Model Engineer* in 1937. At that time, 2.5in gauge track was a popular gauge and LBSC was presenting a modern design. It had the GWR 'look' and the valve gear replicated the prototype.

Don Ashton discovered that this was the point where LBSC's careful interpretation stopped. LBSC freely admitted that valve gear was not his strongest area of understanding but went to the trouble of showing how to machine the driving axle with valve gear and feed pump eccentrics from the solid. It was impossible to get a positive lead on the front port of the valve at any setting of the reverse lever with this arrangement. The problem was due to an insufficient angle of advance on the eccentric to achieve the amount of port opening needed.

Using Charlie Dockstader's 'Stephenson GWR Outside-Admission' simulation programme, you discover that a full 28 degrees of advance and a back-set of 0.142 inches on the expansion link pivot are necessary. A reduction in length of the reverse lever lifting arm to 0.625in is also necessary to get the link suspended from the optimum position.

Little needs to be done to the slide valve other than omitting the exhaust clearance. Don Ashton's version of *Purley Grange* achieves a valve event symmetry score of 97.8%, something that would be hard to beat in any locomotive, model or full-size.

All the revised valve gear dimensions are shown on the accompanying drawing with analysis results in Appendix 2.1

6.1.2 L.M.S Class Princess Royal 4-6-2 by LBSC

Modifications to the Outside Admission Walschaert's Valve Gear

LBSC's design for a 2.5in gauge version of *Princess Royal* was published in the *English Mechanics* in 1933. The design featured a four-cylinder layout with the outside valve gear driving the valves above the two inside cylinders via rockers. The software used is not capable of simulating the valve events of the inside cylinders and it must be emphasised that the rocking levers cannot be relied upon to replicate the valve events of the outside cylinders. Don Ashton emphasised this and advised model engineers to avoid using rocking levers wherever possible.

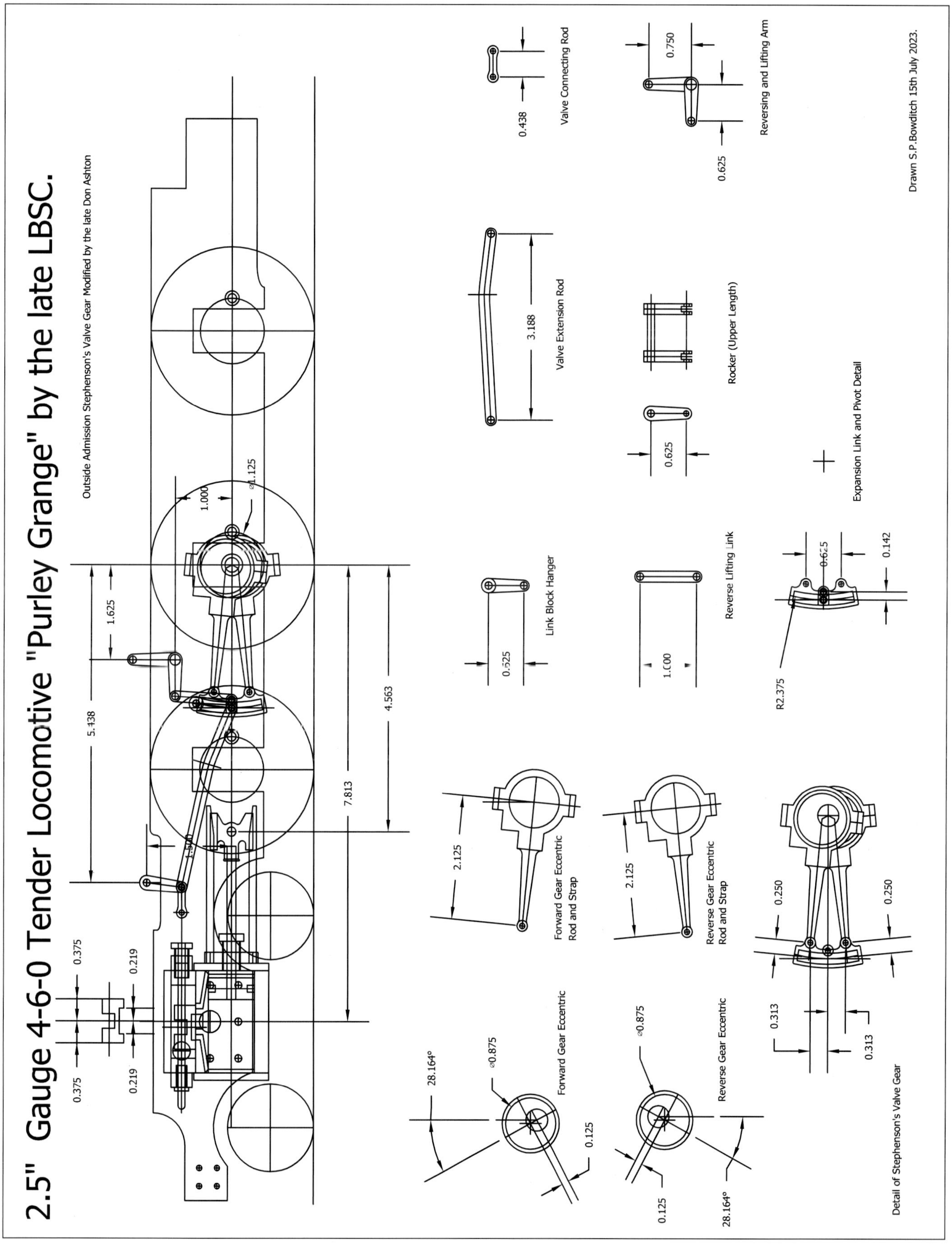

Modified Valve Gear for *Purley Grange.*

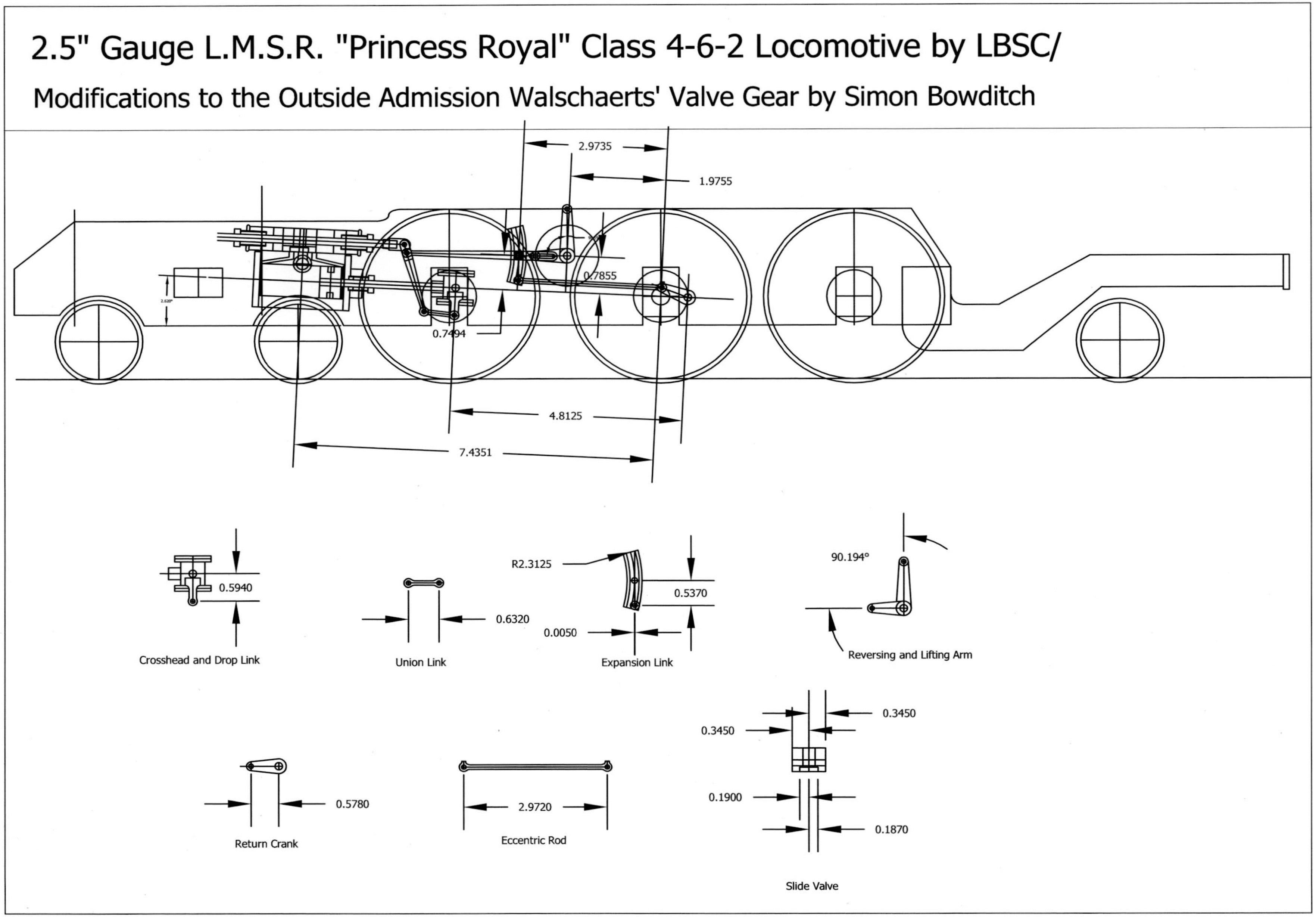

Modified Valve Gear for *Princess Royal*.

The layout of the outside admission Walschaert's Valve Gear is modelled to give a close appearance to that of the full-size locomotive except for the point of attachment of the radius rod to the combination lever. This is because the full-size locomotive used inside admission piston valves, whereas the model uses outside admission slide valves, necessitating the pivots points for the radius rod and valve spindle to be swapped.

The alteration to the angle of inclination of the radius rod caused by this arrangement means that the location of the vertical centre of the expansion link needs to be optimised and subsequently the vertical position of the weighshaft needs a similar adjustment. LBSC used various strategies to optimise his valve gears and an initial simulation shows that the valve events in forward gear are very good. However, events in reverse gear are less impressive, scoring 82.3% for symmetry. This may not be an issue for an express engine, but addressing the issue might improve the overall valve event symmetry score.

Using Charlie Dockstader's outside admission Walschaerts' simulation program, I found that some improvement could be obtained by making the following adjustments.

Firstly, the location of the vertical centre of the expansion link was increased by 0.078in and that of the weighshaft was reduced by 0.036in. Next the length of the drop link attached to the crosshead was increased by 0.031in and the length of the union link was increased by 0.007in. LBSC gave no clear location for the location of the driving pin on the expansion link. I have redrawn this component to have a vertical distance of 0.573in from the centre of the trunnion to the horizontal centre line of the driving pin. The 'offset' figure for the centre of the pin is just 0.005in off the vertical centre.

The diameter of the eccentric crank circle remains the same, but a new slightly shorter return crank is needed as is a new 'optimised' eccentric rod. Finally, if you want to be pedantic, the angle between the reverse arm and the lifting arm on the weighshaft should be set at 90.194 degrees. I altered the proportions of the slide valve to marginally increase the outside lap and then proportioned the exhaust cavity to give asymmetric laps to improve the exhaust timing.

The overall improvement gives a valve event symmetry score of 89.7%, representing a gain of 7.4%.

All the revised dimensions are shown on the accompanying drawing with analysis results in Appendix 2.2

6.2 3.5in Gauge Locomotive Valve Gear Analysis

6.2.1 3.5in Gauge Hunslet Narrow Gauge 0-4-0 Saddle Tank *Lilla* by Ross Harrison Modifications to the Outside Admission Stephenson's Valve Gear

Lilla is a model of the classic Large Quarry Hunslet narrow gauge saddle tank. It has outside frames with outside cylinders but with the valve gear and steam chest located inside the frames. The original full size Hunslet locomotive used Stephenson's outside admission valve gear with a locomotive type of expansion link suspended from the reverse eccentric rod pin. As this is not an arrangement conducive to good valve events, Ross Harrison substituted this for a centrally suspended launch link. Unfortunately, when the valve gear designed for the model was simulated using Professor Hall's software, some issues emerged. These included a sub-optimal figure for the angle of advance of the eccentrics and poor proportioning of the suspension linkage of the expansion link, resulting in link-slip.

Charlie Dockstader's outside admission Stephenson's Valve Gear simulation was then used to investigate the performance of the published design and the results are tabulated in '*Lilla* Table One'. It doesn't take long to appreciate the disappointing

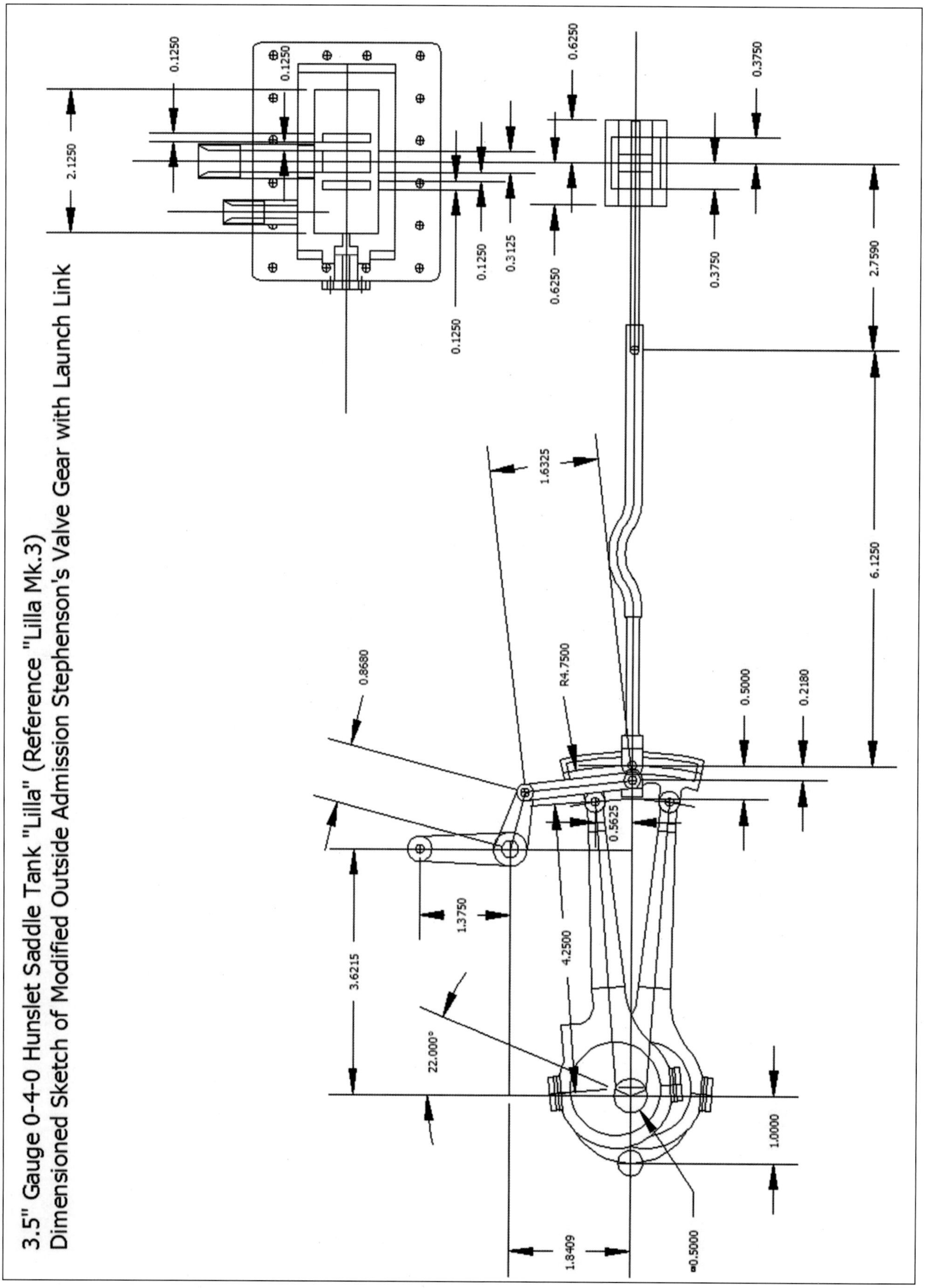

Modified Valve Gear for *Lilla*.

lack of symmetry with an overall low score of 71%.

Lilla Mk 3 is an attempt to show what can be achieved with a correctly proportioned launch link. It will be some relief to anyone attempting a rebuild of an existing 3.5in gauge *Lilla* that the existing eccentrics and eccentric rods can still be used but the original setting, giving an angle of advance of 29.3 degrees, must be retimed to an angle of 22.0 degrees. The components needing replacement are the expansion link with its suspension point back-set to 0.219in behind the curved vertical centre line of the link. The reversing lever, lifting arm and lifting link all need replacing and the weighshaft is located on new centres on the frame. The result with the launch link compares well with the improvement achieved with the locomotive link with the overall symmetry scoring 94.1%.

Lilla 'Version C' is a rebuild of *Lilla's* outside admission Stephenson Valve Gear with a centrally suspended locomotive link. This modification is quite extensive in that it requires new eccentric sheaves with a throw of 5⁄16 in which must be set to give an angle of advance of 15.7 degrees. It will also need longer eccentric rods and please note that the link is suspended from a point 0.063in FORWARD of the curved centre line of that component. There are alterations to the length of the reversing lever, lifting arm and lifting link. Thankfully, the only alteration to the cylinder assembly is to reduce the valve length by 0.010in on both leading faces.

The overall result is much improved, scoring 98% for symmetry and leads and port opens on both front and back port coming very close to the accepted ideal of being equal.

The cylinder clearance volumes on *Lilla* might appear to be large, because of the very long steam passages between port face and the cylinder ends. When this was calculated, it was found to be only 6.562%. It is therefore clear that *Lilla* will not be tolerant of inaccurate valve timings and that care must be exercised during this critical process.

All the revised dimensions are shown on the accompanying drawing with analysis results in Appendix 2.3

6.2.2 3.5in Gauge BR Class 7MT 4-6-2 Britannia by LBSC. Modifications to the Walschaert's Valve Gear

I read somewhere that LBSC's 3.5in gauge *Britannia* had a problem with the valve gear. The simulation does provide an answer to the issues.

The first task was to create a new AutoCAD drawing for the published frame, cylinder, and valve gear drawing to obtain the dimensions essential for simulation. These were not given in either the original articles or in subsequently marketed plans.

Using Professor Hall's Walschaert's software requires 21 dimensions from the design drawing to be entered into the data entry window. As well as simulating the valve events produced by the valve gear, the programme will optimise the proportions for the length of the eccentric rod and return crank. In addition, the programme sets the valves to give equal amounts of 'lead' at both the front and rear port when the main crank is on front or back dead centre.

The results indicated that the valve gear provided insufficient motion to the valve to obtain the lead steam necessary for the admission phase of the steam cycle. This would have resulted in the engine being an uncertain starter and sluggish in its overall performance. With no conventional values found in the admission phase the overall valve event symmetry score was a meagre 61.4%.

The lap of the valve was reduced from 0.125in to 0.110in to increase the Lead value to 0.015in. This ensured appropriate and symmetrical values for the admission phase of the steam cycle in all positions of the lifting arm. It still left an unconventional value to the exhaust steam phase of the steam cycle as LBSC had provided an exhaust clearance to the piston valve of 0.015in.

The simulation results from the Dockstader software confirmed that the

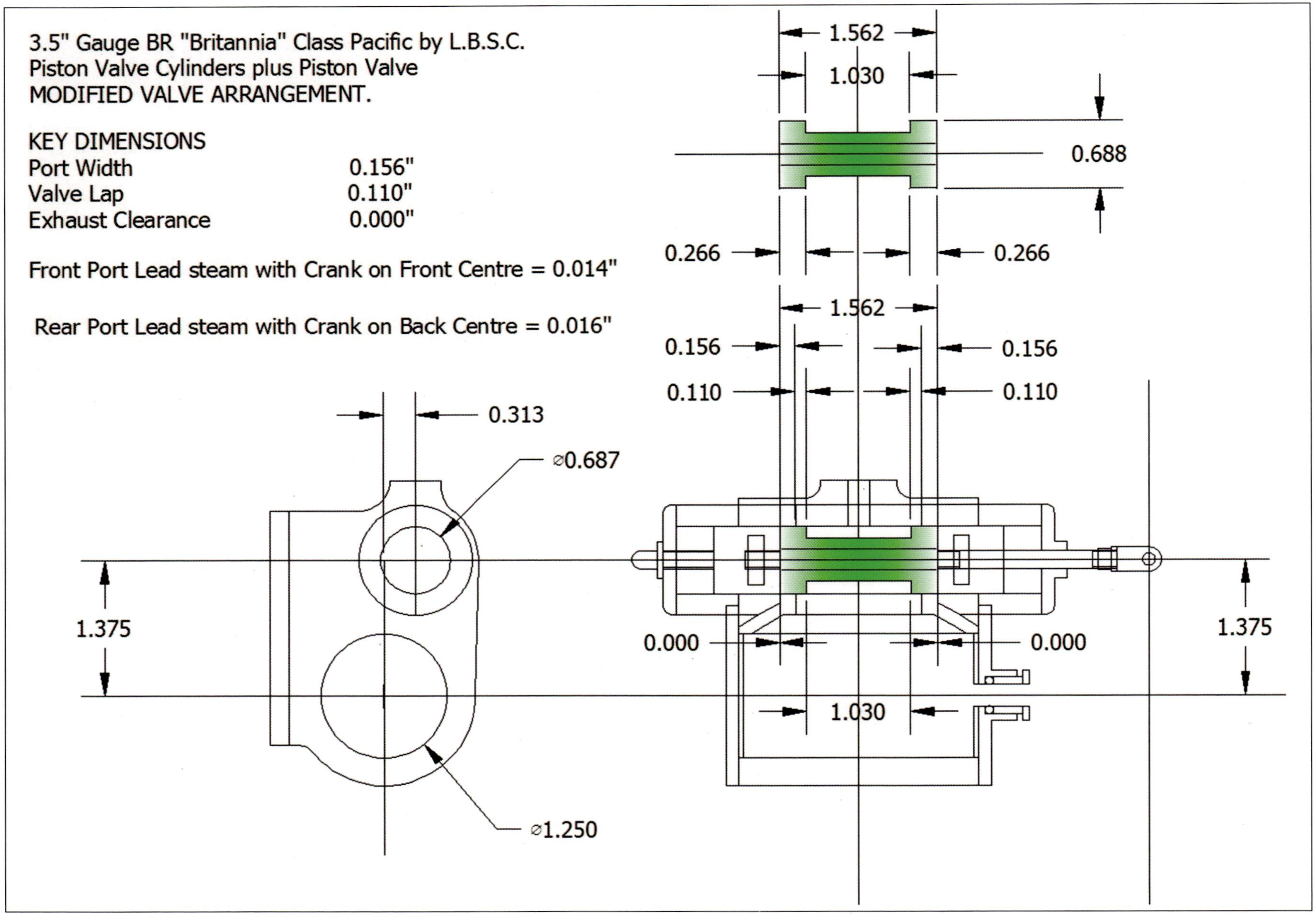

Above and opposite: Modified Valve Gear for *Britannia*.

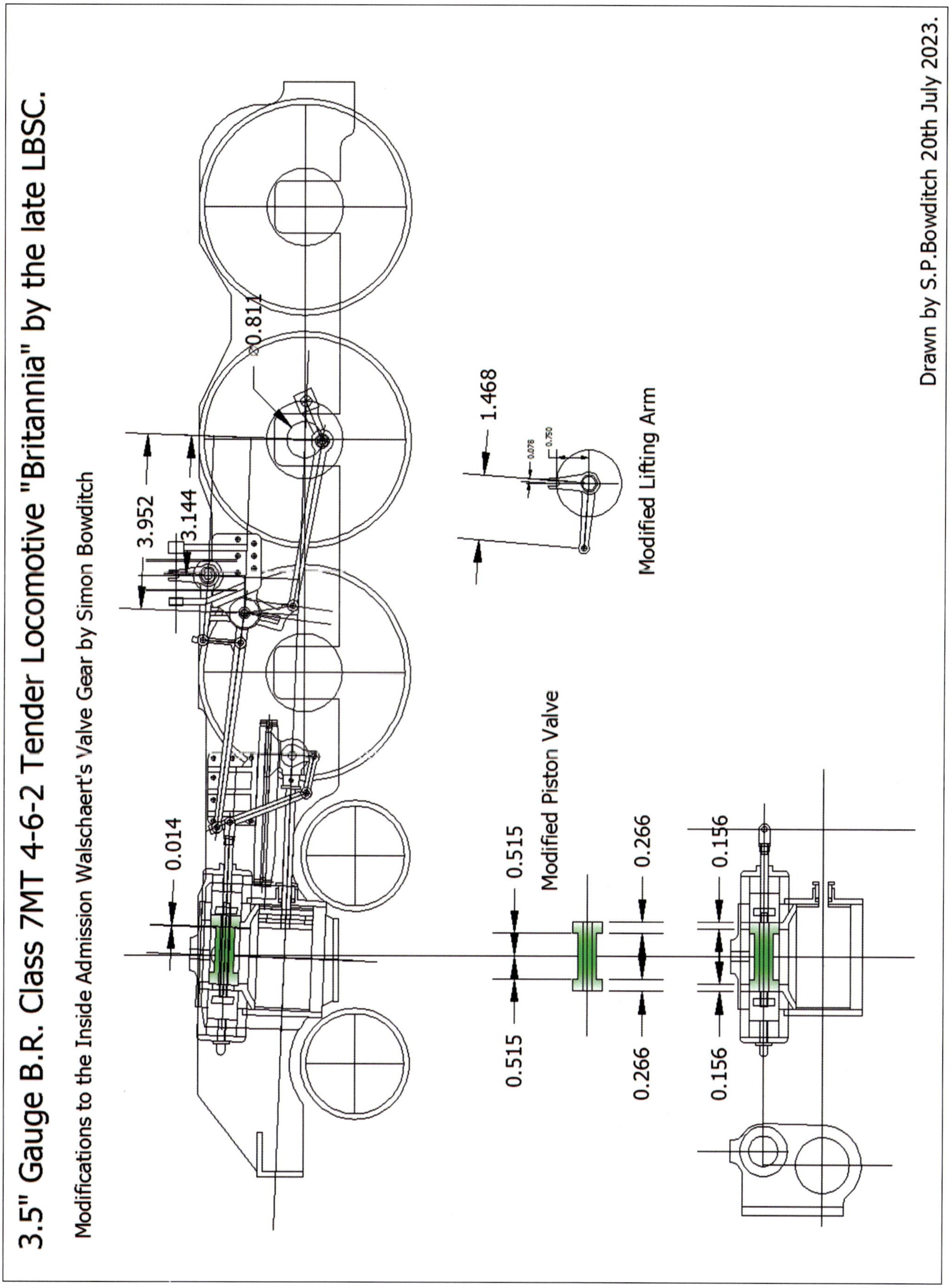
3.5" Gauge B.R. Class 7MT 4-6-2 Tender Locomotive "Britannia" by the late LBSC.
Modifications to the Inside Admission Walschaert's Valve Gear by Simon Bowditch
Drawn by S.P.Bowditch 20th July 2023.
Modified Lifting Arm
Modified Piston Valve
0.811
1.468
3.952
3.144
0.078
0.750
0.014
0.515
0.515
0.266
0.266
0.156
0.156

issue required two new valve chambers or changing the proportions of the valve. By increasing the dimension of Data Entry No.41, valve centre to inside edge of inside ring, from 0.500in to 0.515in, it was possible to obtain a lead value of 0.015in. The unnecessary exhaust clearance of 0.015in on the outside edge of each valve was removed and the valve width of 2.66in retained.

A further improvement to the symmetry of the longer valve travel settings in reverse gear was achieved by reducing Data Entry No. 20, The length of the lifting arm on the weighshaft from 1.500in to 1.468in. The adjustments to the proportions of the original LBSC layout result in an overall improvement in valve event symmetry of 29.4%, with 5.2% of the final percentage score coming from the added sophistication of the Dockstader program. The cut-off values in full reverse gear remain significantly asymmetric but how much time would the locomotive have to spend running in this direction to make a complete redesign worthwhile?

All the revised dimensions are shown on the accompanying drawing with analysis results in Appendix 2.4

6.3 5in Gauge Locomotive Valve Gear Analysis

6.3.1 'Simplex' 0-6-0 Tank Locomotive Valve Gear: Outside Admission Walschaert's

This design was published in 1967 and remains a popular beginner's engine. The ashpan is very shallow where it passes over the rear axle, and this can lead to poor steaming, but is easily remedied by drilling a few holes to admit more air. The valve gear was always thought to be alright in terms of the fact that it could run quite well with the reversing lever notched back to below 40% cut-off and therefore nobody published any adverse comment on this part of the design.

Don Ashton showed how to make a significant improvement in the valve events imparted by the linkage as originally designed.

Using Charlie Dockstader's Outside Admission Walschaert's Valve Gear Simulation software, it is evident that the symmetry of the events of the original design is good enough in places but not across the whole range of the reversing lever.

Don Ashton's modifications to *Simplex*'s valve gear goes a long way towards addressing these issues and involves remaking eight pairs of components of the Walschaert's linkage and a modification to the slide valves.

My Dockstader Input Table for *Simplex* has all the original dimensions listed in the first column and the entire revised dimensions in the second column, highlighted in red for clarity. To assist your understanding of what will be needed, a drawing detailing the revised centres of the various components of the Walschaert's linkage is also included.

The original Valve Event Chart for *Simplex* gave an overall symmetry score of 84.6% across the entire range of reverse lever settings with the least symmetric events being in reverse gear. By the standards of designs of 56 years ago, this was quite good. Now armed with the insights which Don Ashton freely shared with the model engineering community, the results have been improved and the symmetry score becomes 95.9%. This is an improvement of 11.3% in valve event symmetry with most of the gain being in reverse gear. We also see a constant lead value across the range and a smoother "feel" to the engine's performance.

All the revised dimensions are shown on the accompanying drawing with analysis results in Appendix 2.5

6.3.2 Speedy 0-6-0 Tank Locomotive by LBSC Improvements undertaken by Professor Hall

LBSC presented this design in *English Mechanics* magazine in 1950, shortly after the full-sized engines appeared. It had a close to prototype appearance, but it

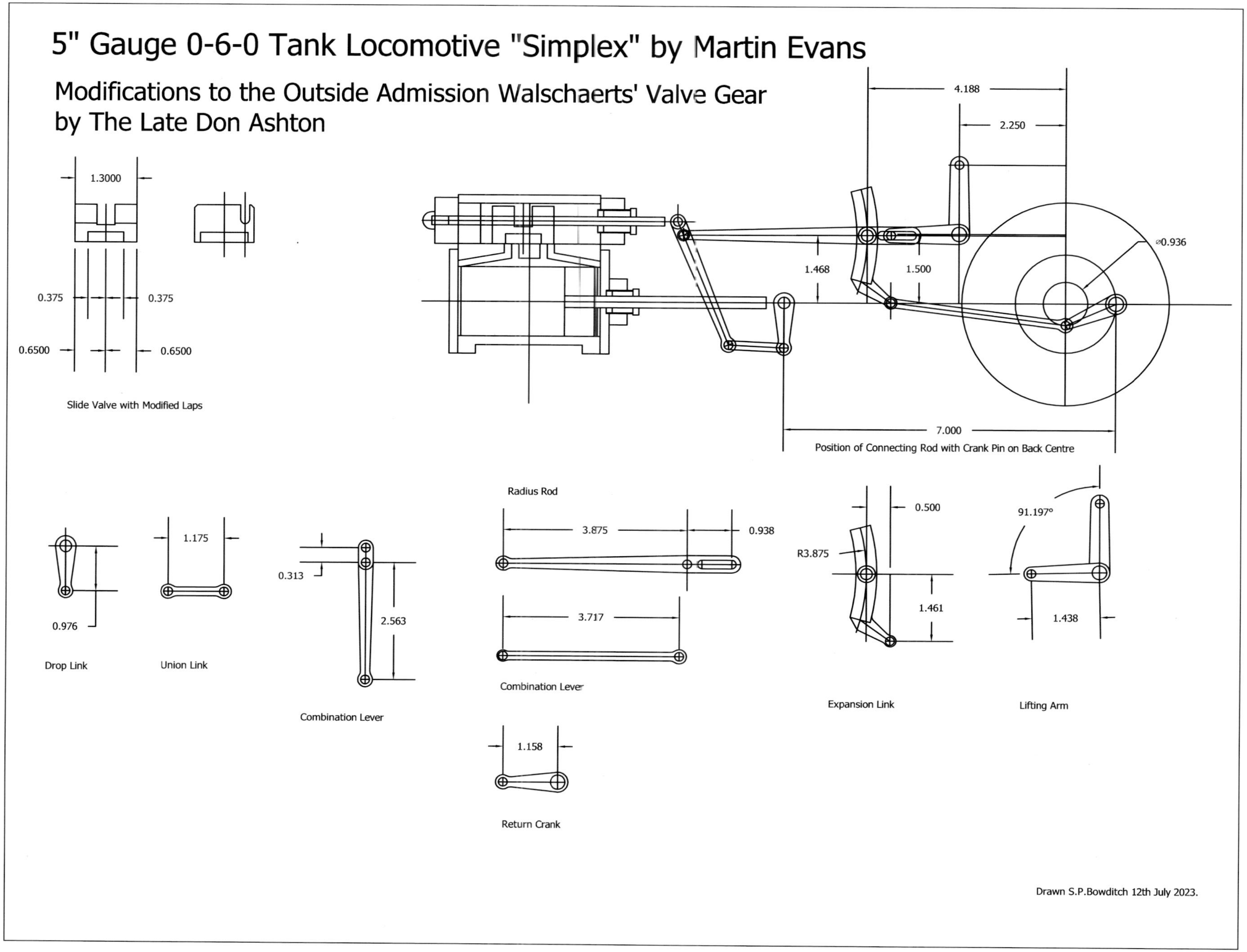

Modified Valve Gear for '*Simplex*'

became evident that all was not well with this engine.

The original design cut-off in full forward gear was discovered to be limited to 50% and both K.N. Harris and Don Young presented redesigns but with little improvements. Don Ashton said that *Speedy*'s valve gear layout suffered from what he called 'compaction' – a common problem with six-coupled engines with short wheelbases.

Professor Hall used his simulator to assess the valve gear and compared it to the full-size proportions. The full-size locomotives had a port width of 1.375in and a lap of 1.250in, similar to that of the 45XX 2-6-2 Tanks. LBSC chose a port width of 0.1875in and lap of 0.1563in. These values when scaled up to full size represent a port of 2.131in and a lap of 1.776in, equating to a port width 155% over scale and lap value equal to 142% of scale value. Clearly LBSC expected his valve gear to achieve a significant amount of travel with little thought for the distorting effects this might have on valve event symmetry.

Professor Hall selected a port width of 0.150in (+ 124% of scale value) and a lap of 0.125in (+113% of Scale Value) and this produced a full forward gear cut-off of 70% and an acceptable lead 0.020in. The improvement in valve event symmetry is significant with an overall score of 91.2%. This represents a 16.9% gain over LBSC's original design.

All the revised dimensions are shown on the accompanying drawing with analysis results in Appendix 2.6.

6.3.3 GWR 48XX / BR 14XX 0-4-2 Tank Locomotive Didcot by Neville Evans with modified Outside Admission Stephenson's Valve Gear by Don Ashton

The outside admission Stephenson's valve gear applied by Neville Evans to his design *Didcot* is closely based on the layout designed by Martin Evans for his Great Western 2-4-0 tank locomotive named *Metro*. Both *Didcot* and *Metro* use outside admission Stephenson's valve gear with the expansion link suspended centrally and the weighshaft mounted above the gear. There is an important difference between the gear used in both these model locomotives and the layout of the original full-size Great Western locomotives. Both the *Metro* Tank and the 48XX/14XX used the arrangement with the expansion link pivoting from the top pin where the forward gear eccentric rod is attached to a lifting link. This was raised and lowered by a lifting arm attached to an underslung weighshaft. This arrangement of the Stephenson gear first appeared in the 1860's.

Unfortunately, some mistakes were made in obtaining the correct angle of advance for the eccentrics, the optimum horizontal back-set for the point of suspension of the expansion link and the distance between its top and bottom Eccentric Rod bearing centres. There are also issues surrounding the optimum length of the Lifting Arm which in the case of *Didcot* results in a valve event symmetry score of only 66.9%.

I suspect that Martin Evans considered that a centrally suspended expansion link would give far more accurate valve events than that found in the GWR prototype arrangement. To a certain extent he would be correct as the valve events on the top pin suspended version of Stephenson's Valve Gear are known to lose some symmetry in cut-off figures from the mid-gear point into the early reverse gear settings. Perhaps for this reason Martin Evans chose to use a centrally suspended expansion link, unaware that the forward gear events of the GWR arrangement were almost perfect.

Don Ashton revisited the whole layout and proposed that several key components of the valve gear be replaced. The eccentrics will need rotating to the correct angle of advance and the expansion links, four eccentric rods, two lifting arms, two lifting links, the reversing arm and two slide valves will also need replacing.

The revised valve gear gives a 25.4% improvement in valve event symmetry, significantly improving its performance on the track.

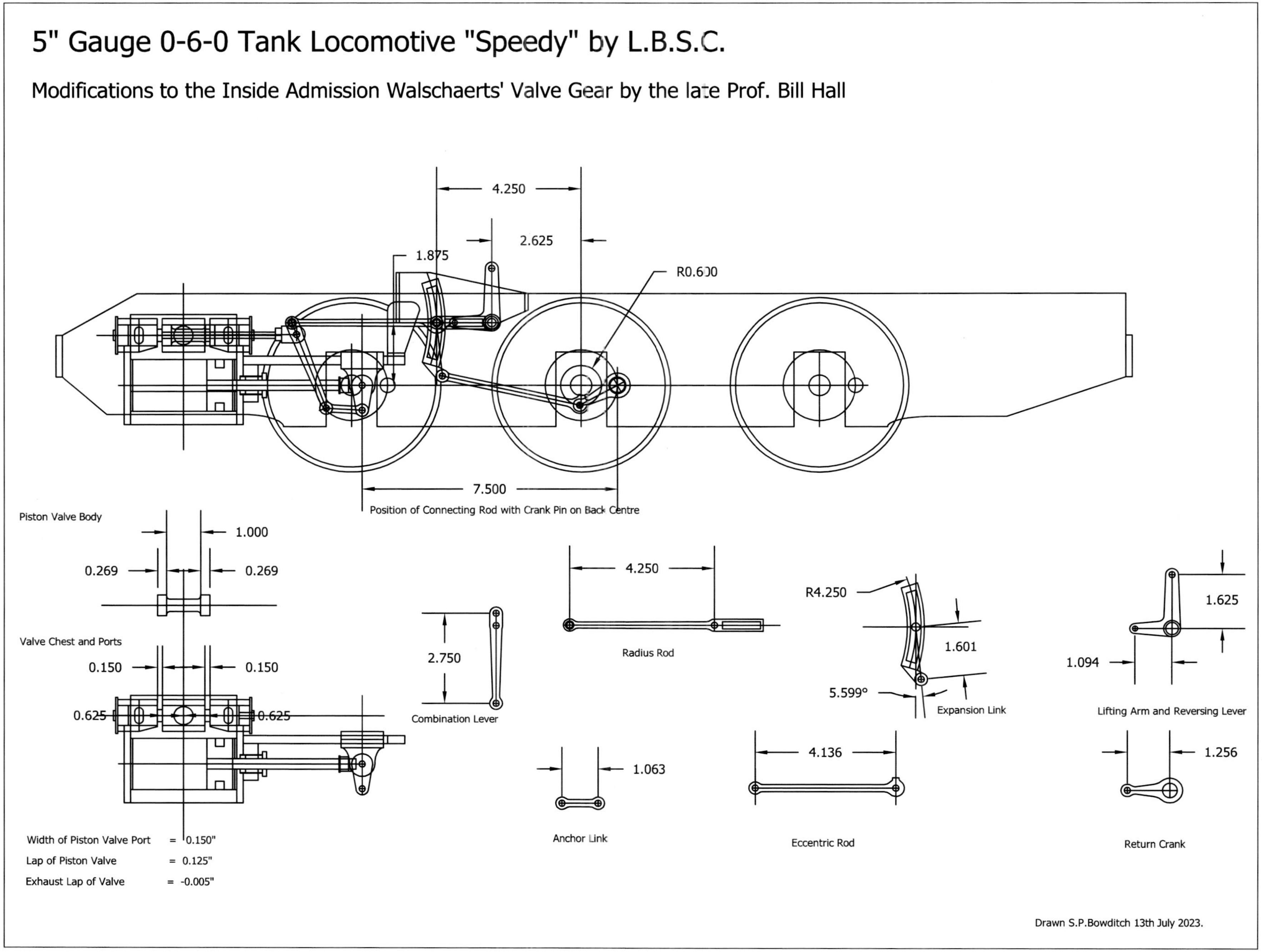

Modified Valve Gear for *Speedy*.

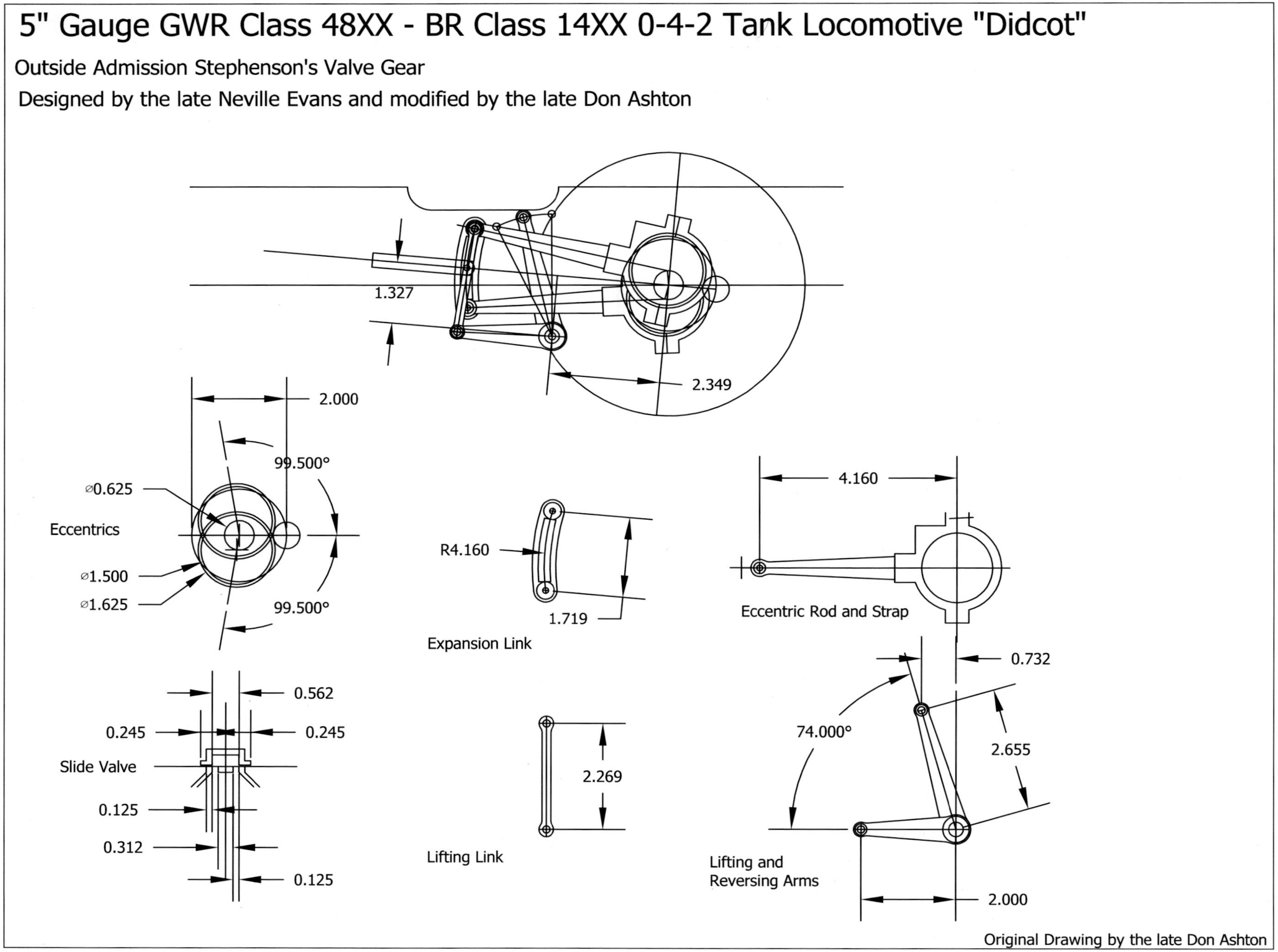

Modified Valve Gear for *Didcot*.

All the revised dimensions are shown on the accompanying drawing with analysis results in Appendix 2.7.

6.3.4 BR Class 2MT 2-6-0 by Don Young. Modifications to the Inside Admission Walschaerts Valve Gear

Don Ashton would have immediately criticised this design for having a combination lever with proportions distorted purely to satisfy the obsession with having a valve lap as long as the port. Although this improves the valve's sealing properties, it results in a significant loss of valve event symmetry as the die block is caused to move up and down in the expansion link at mid stroke. In addition, to achieve an acceptable full gear cut-off figure of 78%, the expansion link has to swing through an angle of 58.5° with the likelihood of the expansion link driving pin hitting the nose of the lifting arm as the main crank pin approaches top dead centre. Don's original design still achieves a valve event symmetry figure of 89.3%, which is slightly better than the full-size locomotive which scores 88.1%.

Sorting out this valve gear in simulation took multiple attempts but I believe that I have addressed the problems identified with the original layout.

The first item to tackle was the proportions of the combination lever where the top portion is reduced from 0.563in to 0.375in. The lower portion I similarly reduced from 2.813in to 2.750in. Having done this, the drop link and the union link must be revisited, resulting in the drop link being reduced from 1.000in to 0.938in and the union link increasing from 1.000in to 1.063in.

The piston valve is modified by increasing the distance from the centre of the inside edge of the valve ring from 0.438in to 0.481in. The valve end width thus being reduced from 0.312in to 0.269in at both front and rear which alters the valve lap from 0.156in to 0.113in.

If we are to keep the existing centres for the location of the weighshaft, the position of the link centre pivot must be moved to comply with the dimensions shown in the table. The baseline for the vertical dimension is taken from the centre line of the motion.

The radius rod is reduced from 3.6875in to 3.609in and this new dimension is repeated when machining a new pair of expansion links. The link crank vertical dimension becomes 1.500in and the back-set increases from 0.195in to 0.250in.

This of course means that the eccentric rod and return crank proportions are altered from 4.973in to 4.870in and 1.266in to 1.238in. The diameter of the return crank circle is left the same at 1.125in.

The final alteration is the angle of the lifting arm on the weighshaft from 81° to 82.632° to achieve a vertical alignment of the reversing arm. The modified valve gear gives a full gear cut-off in forward gear of between 78% and 80% with the rest of the values in the four columns of the chart showing the steam cycle conforming to known acceptable values. The valve event symmetry also improves to a new value of 91.9%. All the revised dimensions are shown on the accompanying drawing with analysis results in Appendix 2.8

6.3.5 G.W.R. 78XX "Manor "Class 4-6-0 with Outside Admission by Martin Evans Modifications to the Valve Gear by Don Ashton

Torquay Manor was described in the *Model Engineer* in 1972 and featured both piston and slide valve versions as part of the series. It slowly became noticeable that the piston valve engine outperformed the slide valve version. Don Ashton correctly identified the reason for this and proposed a remedy requiring the least new parts.

Firstly, he corrected the incorrect setting of 28° for the eccentric's angle of advance in the Stephenson's valve gear. Don reduced this angle to 18.5°. He then increased the radial displacement of the lifting pin on the expansion link from 0.032in to 0.155in. It was generally thought that by anything more than about 1⁄16 in of 'back-set' to this pivot invited the phenomena of 'link slip'. He made a

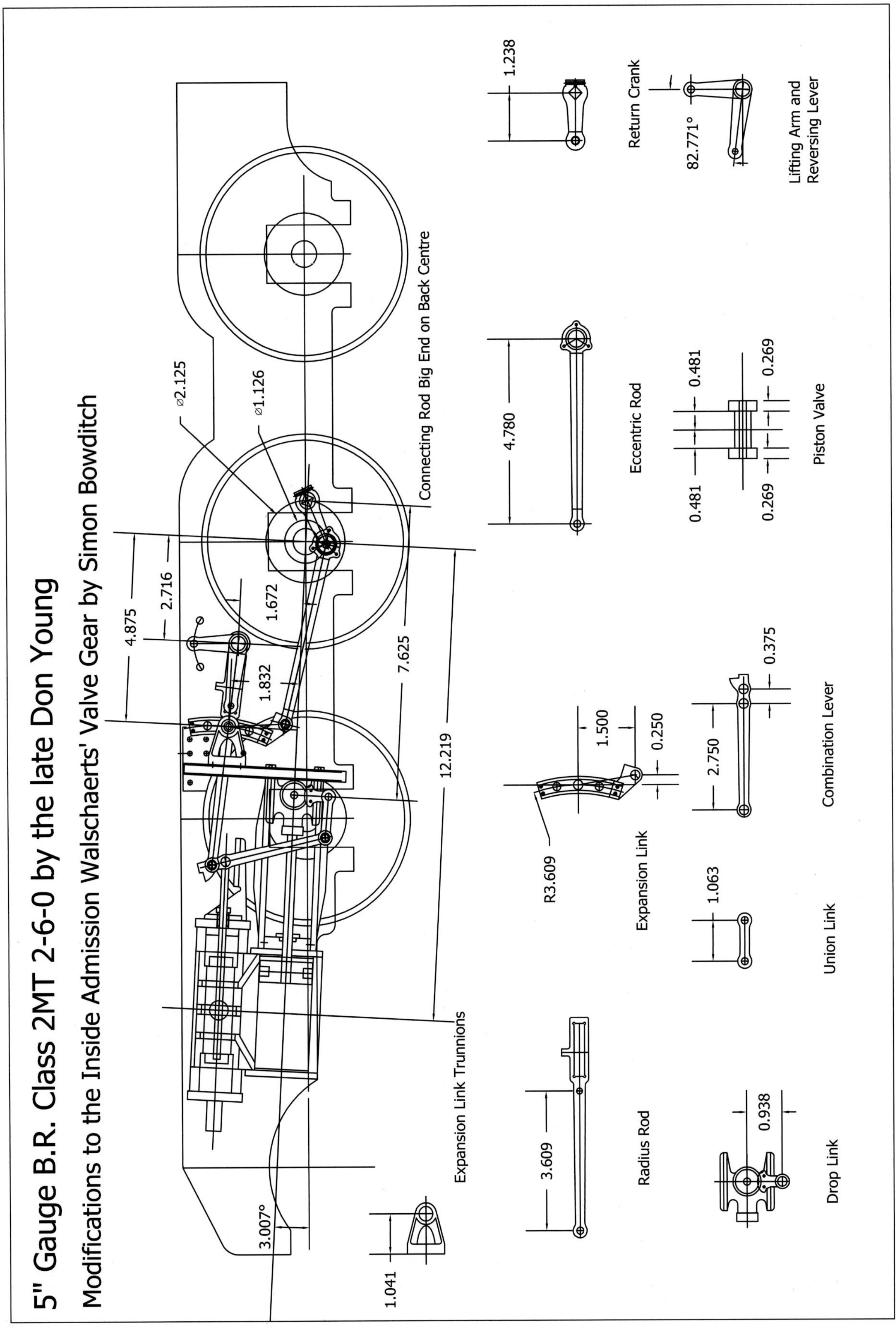

Modified valve gear for 'Class 2MT 2-6-0.

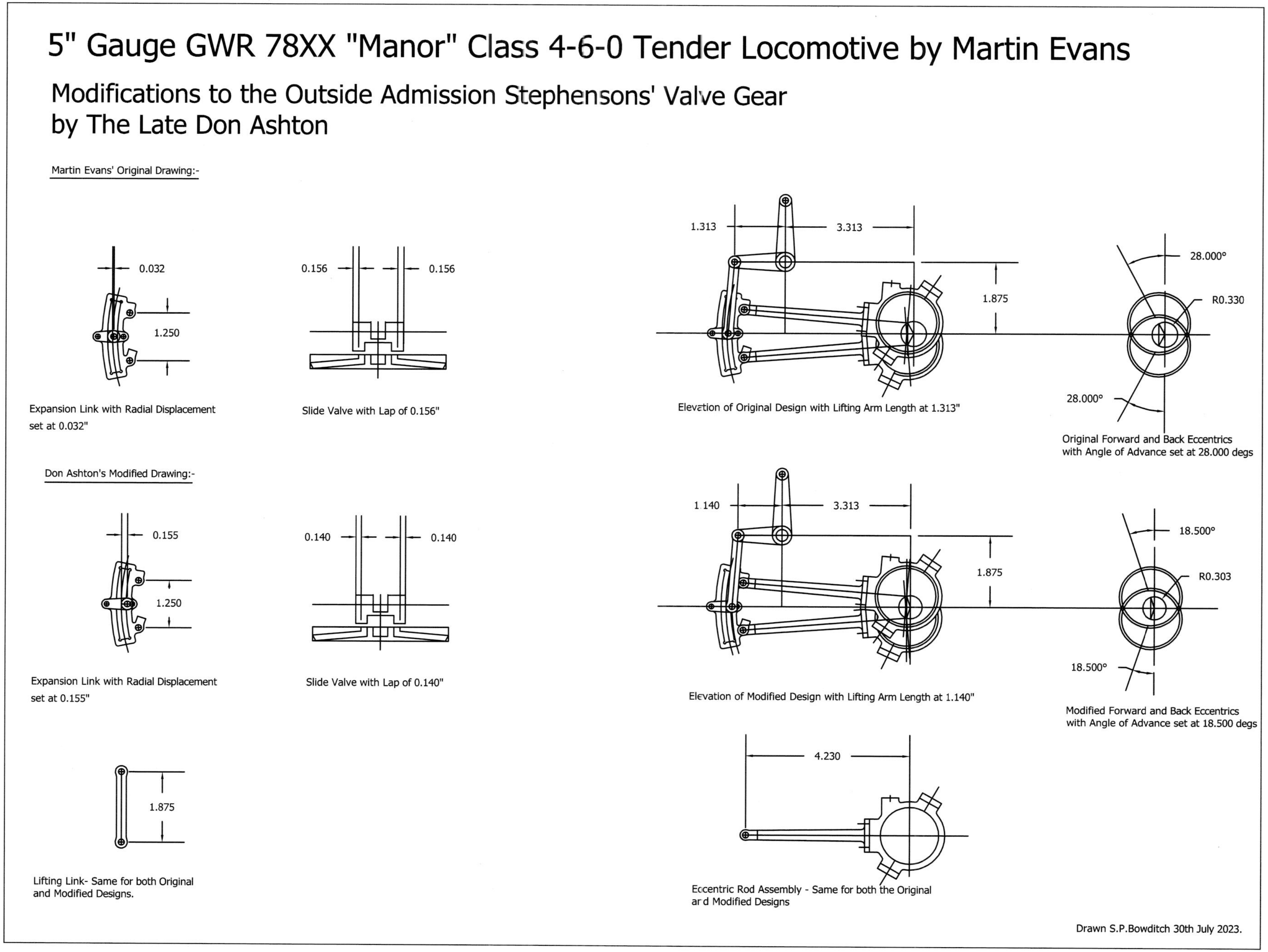

Modified valve gear for *Torquay Manor.'*

further improvement by shortening the lifting arm on the weighshaft from 1.313in to 1.140in. The final alteration was to make a small reduction in the lap of the valve from 0.156in to 0.140in. This was necessary if the existing eccentrics and expansion links were to be retained.

Using Professor Hall's Stephenson valve gear simulator, the original piston valve version delivered a valve event symmetry figure of 67.7%. Unfortunately, the slide valve version achieved only 39.7% – a fact revealed in the Appendix 2 Table 6.3.5.2.

Don could have gone a lot further with this modification but chose to offer his fellow model engineers a solution that was readily achievable in terms of time, material and cost. The valve event chart for the modified version of *Torquay Manor* shows a huge improvement in valve event symmetry achieving a figure of 80.5%.

All the revised dimensions are shown on the accompanying drawing with analysis results in Appendix 2.9

6.4 7.25 inch Gauge Locomotive Valve Gear Analysis

6.4.1 GWR 48XX/BR 14XX Class 0-4-2 Dart by Martin Evans Modified Outside-Admission Stephenson's Valve Gear

The *Dart* design was published in the *Model Engineer* in 1981/82. Again, some issues were noticed with the valve gear, and I was asked to investigate. Using Professor Hall's Stephenson valve gear program, the results showed equal port openings and equal leads across most of the range. However, the valve events symmetry scored only 68.4%. I sought to provide a solution that retained as much of the valve gear as possible, as well as the crank axle, eccentrics, and the location of the bearing/ pivots for the weighshaft. Drawing heavily on the solution which Don Ashton had produced for the 5in gauge *Didcot* design, I finally arrived at my solution – *Dart 3*. The main difference between this and the original design is that the centrally suspended launch type expansion link are replaced with a locomotive link suspended from the top pin, as per full-size prototype. This also required the eccentric rods to be lengthened to a full 6.000in and the lifting link extended to 2.938in. The eccentrics on the crankshaft are retained but the angle of advance is reduced from 23° to 20°. The whole of the original weighshaft with its lifting arm and reversing arm are retained and located on the same location. The outside admission valves in the steam chest are left unaltered but reset to give a mid-gear lead of 0.028in on the back port with the main crank set accurately on back centre.

The lead of the valve increases as the lever is adjusted towards mid gear yet remains equal across the whole range from full forward to full reverse gear. The port openings show the expected variation, with those of the rear port being greater than the front except at the shorter cut-offs where they become equal. Notice again a loss of symmetry in the cut-off results in shorter cut-off settings in reverse gear. This is a characteristic of Stephenson valve gear with the locomotive link suspended by the top pin.

All the revised dimensions are shown on the accompanying drawing with analysis results in Appendix 2.10

It is very pleasing to see that *Dart 3* achieves an improvement in valve event symmetry of 22.8% over the original design, with a final figure of 91.2%.

6.4.2 B.R. Class 2MT 2-6-2 Tank by Dick Stockings Modifications to the Inside Admission Walschaerts Valve Gear

This case study is a follow up on the description of the modifications made to the valve gear of the 5in *BR Class 2MT 2-6-0* tender locomotive designed by Don Young. Once again, we see a designer attempting to make his design benefit from 'long lap-long travel' valves with a valve gear layout which was conceived at a time when the main concern was servicing and repair costs rather than immaculately symmetrical valve events.

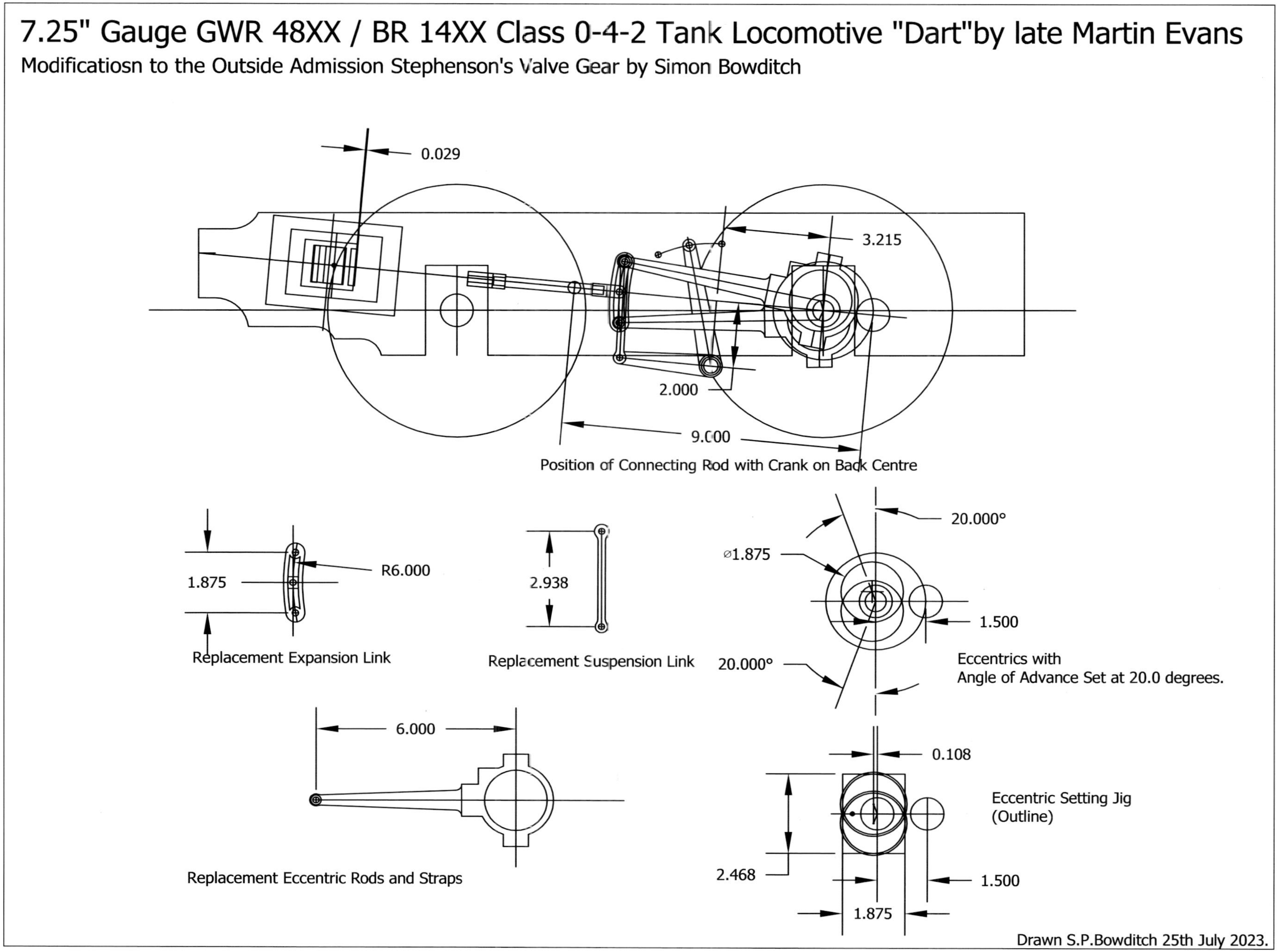

Modified valve gear for *Dart*.

Having the original British Railways drawings for the prototype locomotive available for scrutiny made the job of obtaining the data for a simulation relatively straightforward. The valve event summary on the works drawings, indicated a valve event symmetry figure of 88.1%. The Dockstader simulation of the dimensions taken from the drawing gave a figure of 86.9%. The original measurements were taken by hand with tolerances of +/- 1/32 in whereas Dockstader makes no allowances for the effects of lost motion. A second Dockstader simulation, gave a valve event symmetry figure 81.6%.

The graphic output from the Dockstader simulation program showed the familiar appearance of a valve gear using distorted values to achieve long valve travel. The swing of the expansion link is excessive at nearly 55° and the die block slip in the expansion link is significant because of the over-scale proportioning of the top section of the combination lever. The miniature locomotive uses a piston valve with a 7/32 in lap over a 7/32 in port, whereas the original full-sized locomotive used a 1 5/16 in lap over a 1 3/4 in port. The lap/port proportions of the model valve are 100% of the port width, whereas the full-size locomotive the lap is only 75%.

For the modified version of the miniature gear, the lap is only 77% of the port width. The swing of the expansion link has been reduced to 48.32°. The real answer probably lies in the cylinders, where a reduction in the port width to 3/16 in would help the model locomotive's valve events enormously.

The distance between the two top pivots of the combination lever have been reduced from 0.590in to 0.558in and the pitch circle diameter of the return crank has also been reduced from 1.810in to 1.624in. This alteration requires a shorter return crank and a modification to the length of the eccentric rod. The clearance between the expansion link and the face of the motion plate looked tight on its full forward path, so the location of the eccentric rod pin on the expansion link has been moved. The location of the weighshaft in the frames has also been relocated to give a straight line from the combination lever top pin, through the expansion link centre to the weighshaft centre when the main crank was on front and back centre.

All these modifications have resulted in a slight improvement in the valve event symmetry to 86%. Considering the Dockstader simulation only awarded the full-size locomotive a similar score, we might be getting nearer to an optimal outcome for this locomotive.

All the revised dimensions are shown on the accompanying drawing with analysis results in Appendix 2.11.

6.4.3 LMS Class 5MT 4-6-0 Highlander by Martin Evans. Inside Admission Walschaert's Valve Gear as modified by Don Ashton

Don Ashton expressed concerns for the proportions of the valve gear for this engine and suggested the following alterations so that the locomotive could achieve its full potential. The locations of the horizontal centres of the weighshaft and expansion link trunnions vary only by 0.003–0.004in but the vertical positions are relocated by 0.015in and 0.035in respectively. This achieves an optimum alignment of the radius rod from its point of attachment at the top pin of the combination lever, through the centre of the expansion link trunnion to the die block in the radius rod extension and the centre of the die block on the lifting arm. The front portion of the radius rod is reduced from 9.875in to 9.630in.

While the length of the lower section of the combination lever remains the same, the length of the top section is reduced from 0.625in to 0.602in. The length of the union link is reduced from 1.719in to a more prototypical 1.400in. Likewise, the drop link has been reduced from 1.688in to 1.656in. Optimisation of the lengths of the return crank, return crank pitch circle, eccentric rod and expansion link are now required.

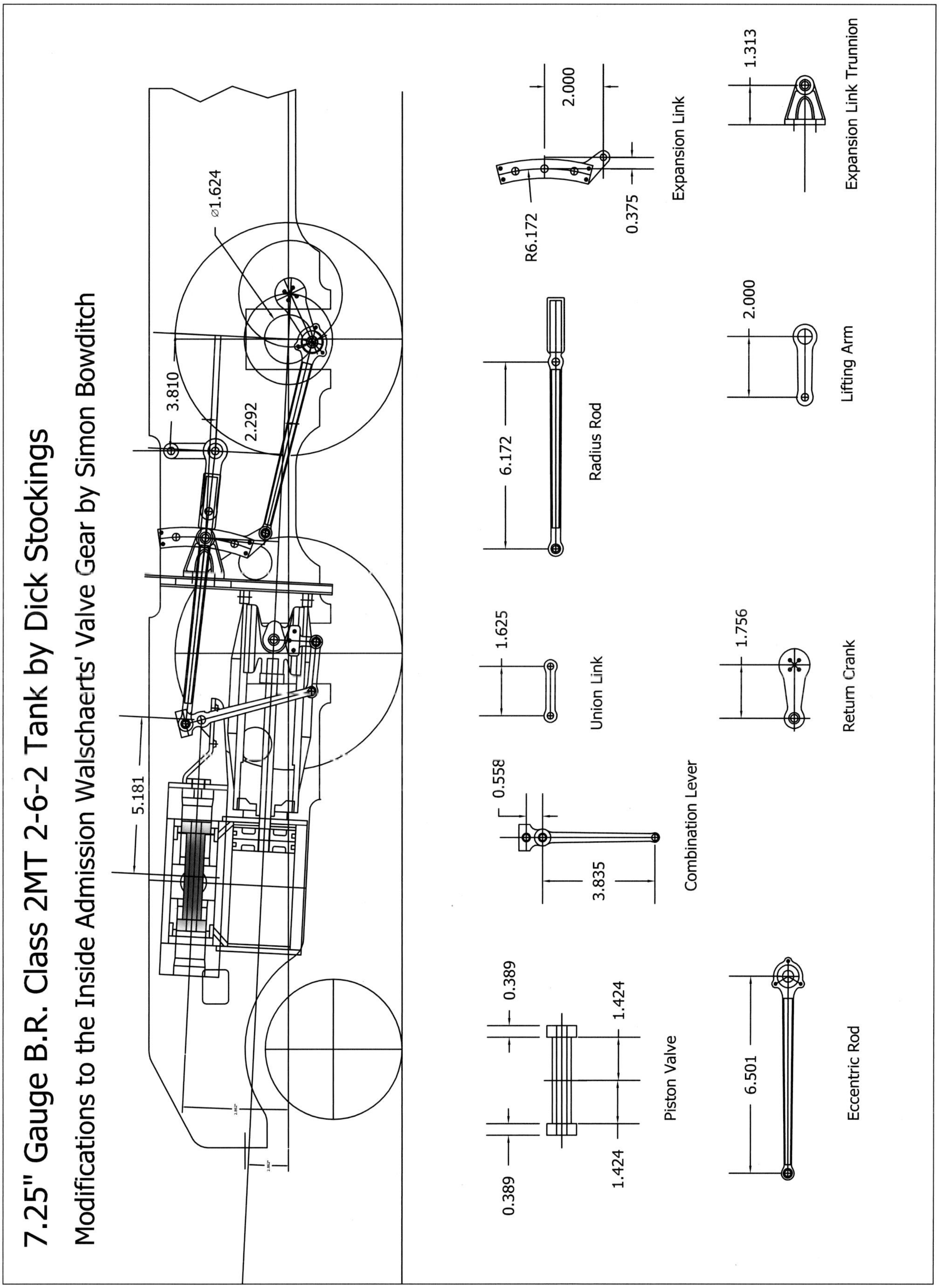

Modified valve gear for BR Class 2 Tank.

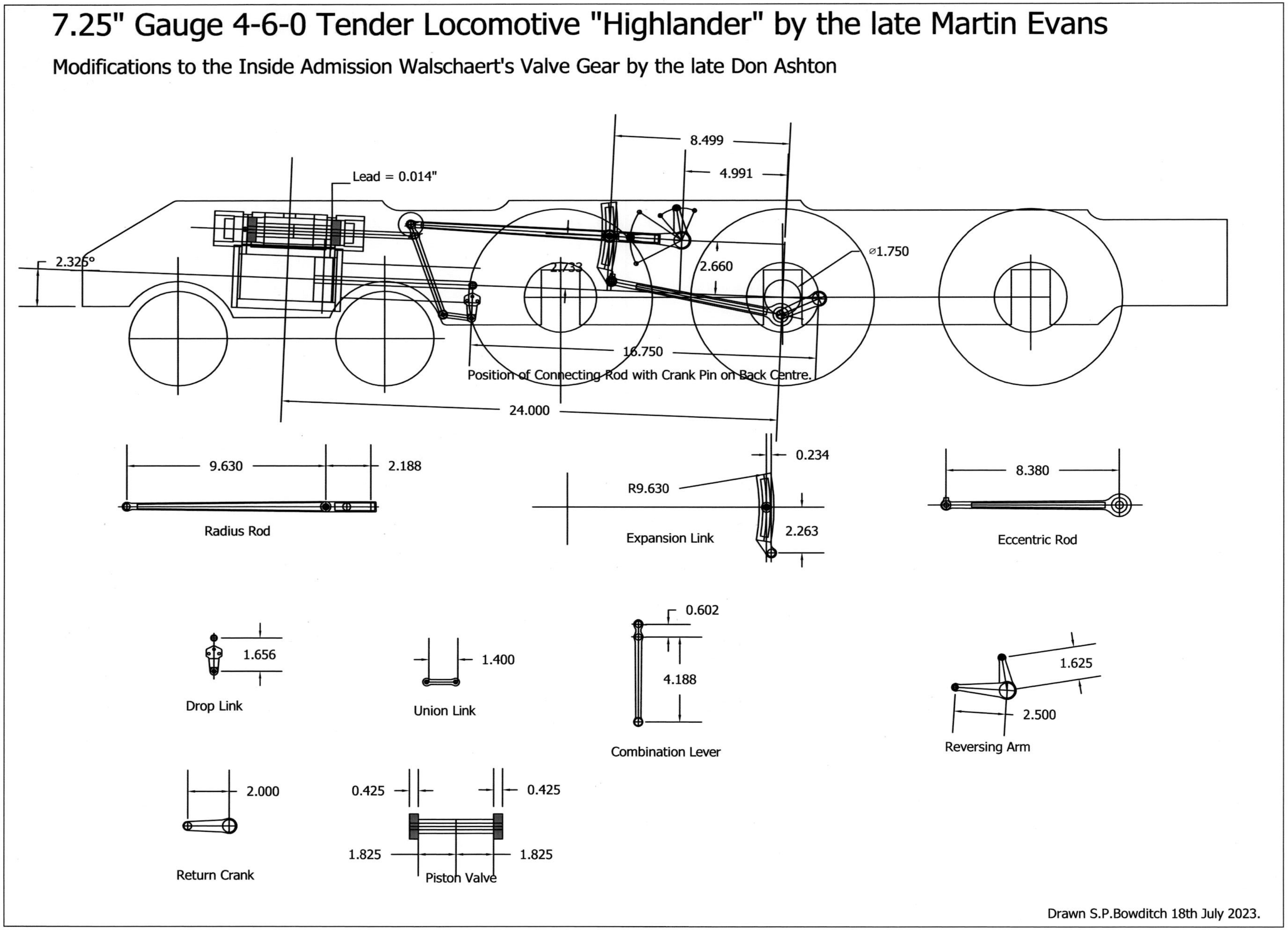

Modified valve Gear for Black 5 *Highlander.*

The return crank length is reduced from 2.032in to 2.000in and the pin now describes a pitch circle of 1.750in diameter. The location of the tail pin of the expansion link is located by the intersection of lines offset vertically 2.263in and horizontally 0.234in from the centre lines of the Trunnion bearing. The eccentric rod is optimised by increasing its length from 8.188in to 8.370in. (This dimension should be checked from your model before machining!)

The existing reversing, vertical arm and lifting arm are left unaltered, but it is now time to look carefully at the dimensions of the piston valve. The position of the ports in the valve chamber remains unaltered.

The distance between the centre point of the two inside faces of the valve-bobbin increases from 1.812in to 1.825in and the width of the valve bobbins is reduced from 0.438in to 0.425in. With the existing port dimensions. this gives a valve lap of 0.206in and a nil value for exhaust lap. The lead should be constant at 0.014in on both front and rear ports.

Don Ashton's modifications to the *Highlander* design show a valve event symmetry score of 94.9%, representing a 13.3% improvement on the values obtained from the original design. All the revised dimensions are shown on the accompanying drawing with analysis results in Appendix 2.12.

CHAPTER 7

STEAM LOCOMOTIVE BOILER DESIGN INFORMATION

7.1 General

The locomotive boiler is the primary source of power for a steam engine and its design must not only meet the demands of the cylinders, but also ensure that it remains completely safe for the pressures it will be working at. Whilst Australian model engineers enjoy a specific code of practice for model locomotive boilers, no such definitive document has yet to be produced for the UK. However, many books, papers and guidelines have been published by a wide variety of engineers and these have been used for published designs for many years without incident.

This work will limit itself to describing conventional coal fired locomotive boilers made of copper with silver soldered joints. It is the material I have used for all my own boilers, and I think that the processes are well suited to the skills of the average model engineer, if advice is taken from a club boiler inspector and the guidance in published works.

Boiler with a flangeless throatplate leaking at the junction with the barrel.

The great thing about copper is that it is so 'user friendly'. With the appropriate sources of heat, it is easily annealed, allowing many shapes and flanges to be formed on the various plates by simply working over formers with frequent annealing.

It is essential to be able to evidence the structural integrity of a pressure vessel by applying the appropriate formulae, calculating the stresses within a joint and therefore determine the factor of safety of the joint.

Early boiler designs often used soft soldered and riveted joints, but as the size and pressures increased, brazing and Sif-bronze joints were often suggested. Currently there is a trend to form joints by copper welding, but this demands the expertise of a specialist welder.

For copper boilers, built by a model engineer, my preference would always to be to use silver solder for all joints.

Some designers have used extra thick tubeplates to avoid using flanges. The benefit is that it allows more tubes to be installed, but I have witnessed failures with this type of joint and would not recommend it.

7.2 Boiler Design Formulae

The following formulae provide a good guide to the thickness of the various parts of the boiler.

7.2.1 The Barrel

$$\text{Thickness of the Barrel } T = \frac{D \times F \times WP}{TS \times R \times Te \times 2}$$

Where

T = Barrel Thickness in inches
D = Boiler Diameter in inches
F = Factor of Safety – Suggested Value = 8
WP = Working Pressure
TS = Tensile Strength (Copper is 25,000 lbs/in^2)
R = Riveting allowance
Te = Temperature allowance

7.2.2 Flanged Plates

These include the smokebox Tubeplate, Firebox Tubeplates, Throatplate and backhead.

Tubeplates should all be the same thickness as the barrel with flanges around the perimeter of minimum depth of ⅜ in.

The firebox outer wrapper should also be the same thickness of the barrel.

The firebox inner wrapper is recommended to be in the order 0.6 to 0.7 of the barrel thickness.

7.2.3 Types of Stays

The following types of stays are used in miniature locomotive boilers: -

7.2.3.1 Side Stays

These should be fitted between any flat surfaces which are typically found around all four sides of the firebox. The pitch of these will depend upon the thickness of the plates, the working pressure of the boiler and the diameter of the stays.

These are often formed by using copper rivets silver soldered with the round heads inside the firebox.

7.2.3.2 Longitudinal Stays

These stays restrain the flat end plates of the boiler connecting the smokebox tubeplate to the backhead. The actual number will depend on the exposed areas of flat plate and the working pressure of the boiler. One or two of the stays are usually hollow for steam connections through to the smokebox i.e. a blower and ejector exhaust. When these are silver soldered at each end, these stays often sag following a hydraulic test. To mitigate against this, these stays can have threaded connections within the smokebox, with a system that allows their tension to be restored.

7.2.3.3 Crown Stays

Many early miniature boiler designs featured the girder type of crown stay but it is difficult to fully connect this type of stay along their whole length, both top and bottom. Failure to achieve full soldered joints on these flanges could leave part of the flat areas of the boiler unsupported, with the inevitable resultant bulging under pressure. A method that seems to have found favour recently is the use of 'rod stays'. These are copper or bronze rods simply passed through both the inner and outer firebox wrappers and silver soldered in place. Again, the pitch is determined by the operating pressure of the boiler, the thickness of the inner firebox plate and the size of the stay. Jim Ewins used his own unique design for crown stays in his experimental 9F and this was presented within the *Engineering in Miniature* magazine series of articles – see Bibliography.

7.2.4 Calculating the Pitch of Stays

The following formula is used to determine the pitch of stays:

Where

$$P = \frac{\sqrt{D^2 \times TS \times 3}}{WP \times F \times 4}$$

P = Pitch of the Stays in inches
D = Min. Diameter of the stay
TS = Ultimate Tensile Strength of the material (25,000lb / in^2 for Copper
WP = Max/ Working Pressure of the Boiler in lb / in^2
F = Factor of Safety - say 8

7.2.5 Number of Smoke Tubes

This is related to the grate area of the engine to ensure good flow of exhaust gases with optimum heat transfer.

It is generally accepted that the total cross-sectional area of the smoke tubes should not be less than $\frac{1}{7}$ but can be up to $\frac{1}{6}$ of the grate area. For Eo = 80, this equates to one tube for every square inch of grate area.

7.2.6 Number of Superheater Flue Tubes

The number of superheater flue tubes is related to the diameter of the cylinders. When calculating the number of smoke tubes required, the flue tubes are counted as smoke tubes, due to the presence of the superheater elements.

7.3 Using Cadmium Free Silver Solder

Brazing is a process to join metallic materials by the melting, capillary flow, and subsequent solidification of a brazing filler metal. The brazing filler metal has a melting range below that of the base materials, making the process distinct from welding processes. Owing to the typically high silver content of most common silver based brazing filler metals, the name of silver soldering is commonplace within industry and amateur metalworkers.

Historically, cadmium containing silver solders have been very popular within the model engineering community. Cadmium has been identified as creating potential health hazards for operatives and national and international bodies have decreed that safer alternatives should be used where they exist. In 2011, Commission Regulation (EU) No 494/2011 dictated that brazing fillers (silver solders) shall not be placed on the market if the concentration of cadmium is equal to or greater than 0.01% by weight.

The banning of cadmium containing silver solders has had an impact on some model engineers, as they have encountered difficulties in making successful joints with the exclusion of this element. Metallurgically, cadmium provides excellent flowability to silver solders which flow beautifully at brazing temperatures. Metallic cadmium melts at 321°C but silver solder typically melts above 650°C and at these higher temperatures, cadmium fumes are liberated, which are both toxic and carcinogenic. As a result, silver solders are now only sold without cadmium and are known as cadmium free silver solder, in line with health and safety guidelines.

7.3.1 Equivalent Products to Silver Solder Containing Cadmium

The equivalent product to Easy-flo™ and Easy-flo™ No.2 and DIN Argo-flo™ is Silver-flo™ 55

7.3.2 Changes in Brazing Techniques

Cadmium-free products typically have higher and longer melting ranges, resulting in a small increase in process times and brazing temperatures. Typically, operators will notice that they need to heat components slightly longer before applying

Table 7.3.1: Temperature Range for the Equivalent Silver Solder Rod.

Original Cadmium Product	Temperature Range °C	Equivalent Cadmium Free Product	Temperature Range °C
Easy-flo™	620-630	Silver-flo™55	630-660
Easy-flo™No.2	608-617	Silver-flo™55	630-660

the filler metal. The filler metal may also flow slightly less freely once molten. The table above shows the slightly higher temperatures for Silver-flo™ 55.

Johnson Matthey are a leading supplier of silver solder materials and their website provides detailed information for all of their products.

7.4 Testing of Locomotive Boilers

Pressure vessels are a safety critical asset and must therefore be checked regularly for integrity and soundness. They must also be evidenced as such, assured, and insured to confirm that it is safe to be used on an engine, possibly for use in public places.

Over the years, the testing regime for steam boilers has varied and so it is important to discuss the current requirements with the boiler inspector of the club or private company who will witness the testing.

The usual method of testing a boiler is to subject it to a hydraulic test to twice its working pressure when new and then

A Pannier Boiler with multiple leaks around the Throatplate.

A typical hydraulic test rig.

1.5 times its working pressure for all subsequent tests. It will then have a steam test up to working pressure to ensure that no joints open and leak at the working pressure and temperature. The frequency of these tests has changed over the years, so agreement with your club boiler inspector will ensure you meet the current requirements.

Safety valves will also be checked at the steam test and the usual method is to build up a large fire, open the blower full and force the fire to its maximum heat output. If the safety valves can control the boiler pressure to within 10% of the specified operating pressure, the valves are passed, but otherwise their throughput discharge capacity will need to be increased. Ideally, a locomotive should have two safety valves, and these can be set at slightly different pressures within 5psi. Early designs often only had small outlet holes above the ball for the steam to escape and these often do not meet this 10% overpressure requirement.

7.5 The Cleaning and Maintenance of a Steam Boiler

As a pressure vessel boils water at the elevated temperature corresponding to its working pressure, impurities within the water are separated out and these build up on the hottest surfaces of the boiler plates. Ideally, we would all enjoy a soft water supply or use 'demineralised water' to mitigate against this problem. For those of us operating engines in hard water areas, such deposits, especially around the tubes on the water side of the firebox front tubeplate, can seriously inhibit the boiler's ability to steam and could eventually lead to a boiler failure.

Full-size locomotive boilers are washed out at regular intervals to maintain their steaming performance and miniature boilers should also undergo this process. The frequency of the 'descaling' will depend on the quality of the water supply and amount of use of the engine, but this process should be done at least annually. Commercial kettle de-scaler is usually sufficiently effective, but I have recently been recommended Fernox DS3 as an effective product.

The normal descaling procedure is to seal all boiler outlets, except for the large dome aperture at the top, and then fill the boiler half to two thirds full of the appropriate mixture of descaling product and warm water. This will initiate effervescence, bubbling and fumes in the boiler. Always do this descaling out in the open to avoid fumes and the inevitable water spillage when flushing through. Once all internal activity has ceased, open the blow down valves and insert a hose into the dome to drive out all residual impurities. Depending on how badly the boiler is scaled, this process may need to be repeated two or even three times. It is essential to give the boiler a very thorough washing out after final draining, to avoid priming when next in steam.

CHAPTER 8

INJECTORS BY RICHARD LINKINS

This chapter is aimed at helping model engineers enjoy a fuller understanding of how these wonderful little devices operate and how to ensure their reliability in service. I have invited Richard to write this chapter on injectors as he is an extremely competent and experienced model engineer who I consider to be an expert on this subject. His knowledge has been gained

Richard Linkins driving his beautiful 78xx Class 2 loco.

by many years of study, experimentation, consultations with other experts and a great commitment to mastering the fluid mechanics principles upon which the operation is based.

Who better to present this chapter? – over to Richard!

8.1 Introduction

Injectors can lead to considerable frustration but, when they work properly, they are the most effective and efficient way of feeding water into a boiler. I have been fascinated by injectors for many years and have made around sixty. I have also tested many injectors made by others.

One of the main problems with miniature injectors is that commercial ones are being sold that vary in quality and have design deficiencies. I suspect that most manufacturers and retailers of commercial injectors do not have a full understanding of how to make a good injector. It takes more time to make a good injector than the manufacturers allow. Linden injectors, which were among the best commercial injectors ever made, were sold for £3.50 in the 1950s which is equivalent to £95 today, whereas most commercial injectors are currently sold for less than £60. I find this frustrating as I do not pretend to be an expert and most of what I know has been gleaned from reading what others have written. Most of this information is freely available. Jim Ewins wrote in 1985, 'Unfortunately manufacturers today do not seem to have the knowledge and ability to achieve the reproducibility necessary to be sure that their products will function as serious boiler feed devices.' This is still the case.

8.2 The History of the Injector

The injector was invented by Henri Giffard and he was granted a patent in 1858. Before this, pumps were the only way to feed a boiler with water. Injectors steadily replaced pumps because they required less maintenance and were more efficient. Over the decades, injectors were improved and made more reliable. One of these improvements was to make them capable of lifting water.

8.3 How Injectors Work

There were three cones in Giffard's injector, the steam cone, the combining cone, and the delivery cone.

In the steam cone, due to a reduction in diameter, the pressure is reduced, and the speed of the fluid increases greatly; the pressure energy is converted to Kinetic energy and the pressure falls. The final divergent part of the cone straightens the flow.

In the combining cone, the steam and water mix, with the steam condensing. The latent heat of vaporisation of the steam is given up to the water as it mixes; the result is water at around 65°C. This jet of warm water then exits the small end of the combining cone. Condensation occurs in the first part of the cone.

The jet of water then enters the delivery cone. As the fluid expands through the delivery cone, its pressure rises as the velocity falls. The exit pressure is higher than the boiler pressure and consequently water is fed into the boiler.

The combining cone is split into a draught tube and a forcing tube, although this feature was not present in Giffard's original injector. This allows the injector to self-start, re-start and lift. When the steam is turned on, its volume is such that it cannot escape through the outlet orifice in the combining cone to the overflow; the combining cone gap allows this escape via a non-return valve (full size injectors sometimes use a flap valve or sliding cone) until condensation is fully established. Once condensation starts, the pressure at the gap drops below atmospheric and the valve shuts, allowing the injector to function.

The steam leaves the steam cone at around 2000 feet per second (fps) and the fluid leaves the combining cone at about 400fps. It takes approximately 50 millionths of a second for the fluid to transit the combining cone.

8.4 Model Injectors

LBSC described injectors in the 1930s, but his descriptions were over simplified and not accurate enough. Unfortunately, this led many people to make injectors that did not work or worked poorly. Commercial injectors were also available, but these also had limited performance.

C.M. Keiller and E.J. Linden experimented with miniature injectors in the 1930s and both achieved good results. Keiller did describe the making of injectors and also set out injector proportions, giving a range for each dimension related to the bore of the delivery cone. I have checked numerous model designs against these proportions, and they all correspond within a small margin.

Linden manufactured injectors but was very reticent about giving details. The myth grew up about the 'Linden secret'; I will discuss this later. Linden injectors were comparatively expensive but undoubtedly very well made and reliable.

Through the years several people have described how to make injectors in the model engineering press. Probably the best of these were Basil Palmer's articles, although his designs were optimised for the higher altitudes found in South Africa and this requires slightly different dimensions as the atmospheric pressure is higher in our country.

8.5 Experiences with Injectors

My early experience with injectors was that many of them worked poorly. I then met Jim Ewins and I was allowed to drive his engine. I was amazed at the reliability of his injectors. When the water valve and steam valves were opened, they always started, even if the steam valve was opened first. This was not a fluke, all of Jim's injectors worked impeccably.

When I was building my loco, I wanted to fit injectors and was fortunate enough to know Fred LaRoche. Fred made injectors which worked well and sold them at a moderate price. I fitted two to my loco and was very satisfied. I had no desire to make my own; I had plenty of other things to keep me busy. Later, Fred's descriptions of making injectors and his advice encouraged me to make my own injectors.

8.6 Fitting Injectors

Incorrect fitting of injectors can lead to poor performance. For example, a 25oz/min injector requires 3⁄16 in thin-walled pipe (the wall thickness must be no more than 22 gauge) for all three connections, although John Cashmore specified 5⁄32 in for the steam supply and a test has confirmed that this is satisfactory. The free-way through the valves and fittings must have at least the same area as the pipe, equivalent to a 9⁄64 in hole or a little larger. Clacks also need to have 1⁄16 in lift. There must be no air leak into the water supply as this disturbs the fluid flow and a filter is needed in the water feed as the water gap around the steam cone is only about 0.006in on a 25oz/min injector, so tiny particles can easily clog it. The radius of pipe bends should be at least three times the pipe diameter. If an elbow is required, then the bore should be opened out to 5⁄32 in. For smaller or larger injectors, the pipe bore should be adjusted proportional to the injector feed rate.

Jim Ewins described water valves and filters. The water valve has a PTFE plug which does not stick and an 'O' ring seal on the spindle to stop air leaks.

The filters are made from thin foam (it only needs to be 6mm thick), as used in chair padding, in a two-part tubular holder, again sealed with an 'O' ring. I have found these very successful. The filter elements need to be changed every few runs but can

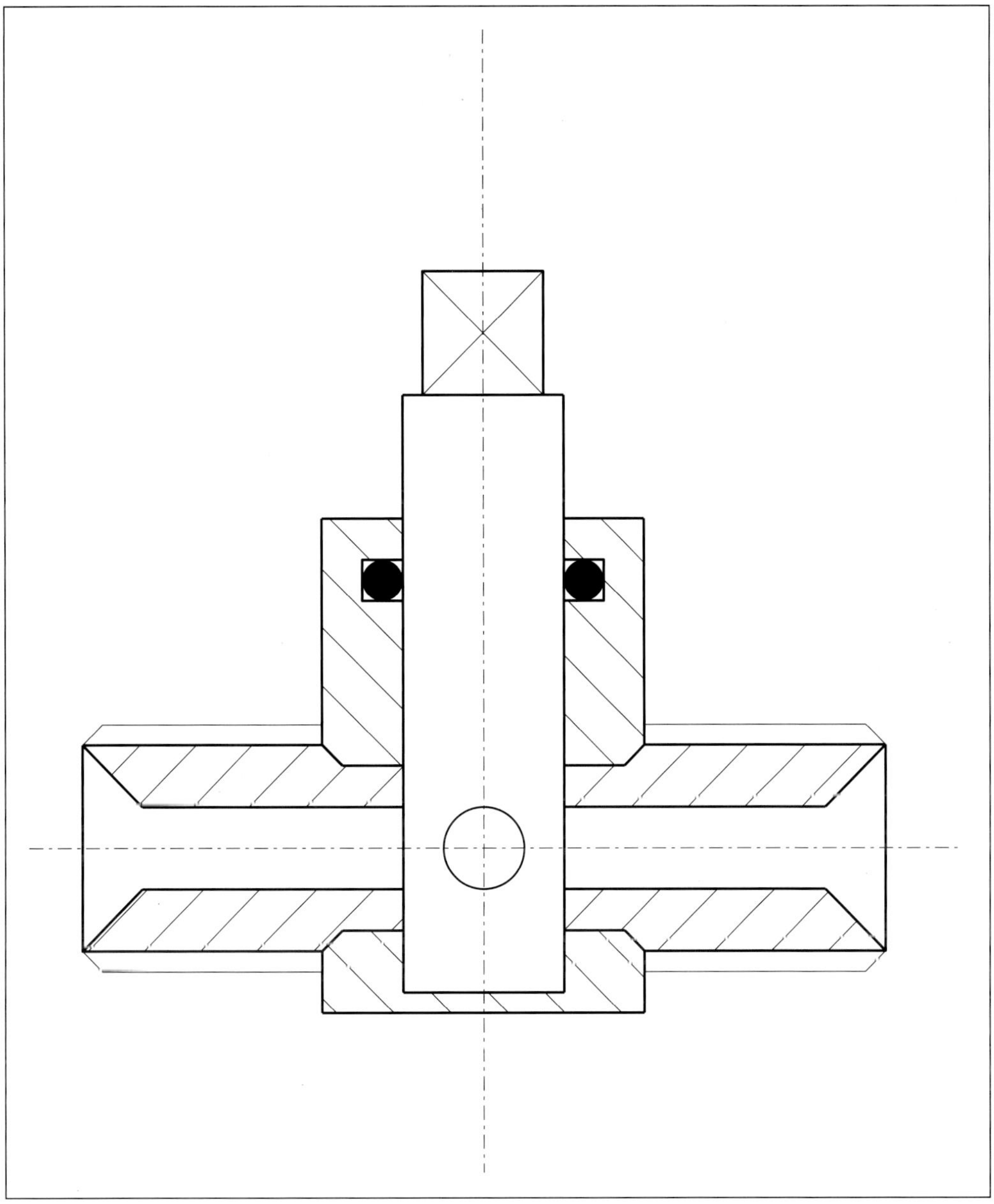

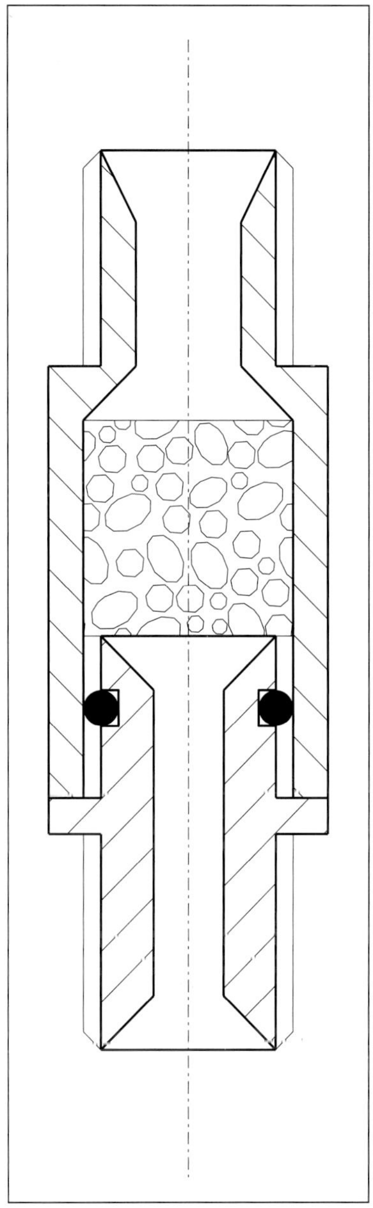

be made from scraps of foam. I once used foam that was too dense; this restricted the water flow and the injector worked poorly. The water should flow from the feed pipe at the same rate as the injector feeds. Restrictions in the water pipework, valve or filter will reduce the maximum working pressure of the injector or stop it working altogether.

8.7 Identifying Problems

It can be difficult to identify problems with injectors fitted to new locos. If there is a problem with the pipework, it is most commonly with the water feed. Connecting the injector to a temporary supply is a good way to check this. All that is needed is a plastic container fitted with a piece of rubber pipe and a stub pipe to fit the injector feed; no valve or filter is necessary for testing. If this does not solve the problem, fitting an injector that is known to work or testing the suspect injector on another loco will narrow the problem down. I carry spare injectors so I can change one that is malfunctioning.

An injector should have 'dry range' of 50 to 60psi. Below this range water will flow

Above left: Jim Ewins design of injector water valve.

Above right: Jim Ewins inline injector water filter.

quickly from the overflow pipe. When this occurs, partially closing the water valve will allow the injector to work down to about 30psi.

If the injector is above its working range, steam and a little water will flow quickly from the overflow. The only way to make it work is to lower the steam pressure or increase the orifice around the steam cone tip; a means of doing this is discussed below.

The flexible pipes should also be checked occasionally as these can split at the ends and cause air leaks.

Taking the water feed from tanks close to the boiler can cause problems if they are not insulated. A separate tank on the back of the engine, in a wagon or on the driving trolley is a much better idea. As noted above, injectors only work with feed up to 40°C. Even tender engines can suffer in hot weather. The first sign of trouble is when the maximum pressure the injector will work at falls. This is because the water is no longer condensing the steam. The same problem can be caused by scale building up around the tip of the steam cone which restricts the water flow. Even full-size hot water injectors only work up to 60°C and they have valves in the overflow to achieve this.

Restrictions in the steam supply will affect the bottom end of the working range first as there will be insufficient energy in the available steam.

If the maximum working pressure of an injector falls over time, it is probably due to scale build up around the steam cone tip, which restricts water flow. Try cleaning the injector immersed in vinegar or kettle descaler but be careful as these can damage the delicate tip of the steam cone if the injector is left in it too long. I use citric acid which takes longer to work but does not seem to damage the tips if the injector is left in it for some time. If this does not work, remove the steam cone and fit a 0.010in shim between the shoulder of the cone and the body. This increases the water gap and raises the maximum working pressure.

When removing cones, treat them with respect. Do not use a big pair of serrated pliers on the shoulder; make a special puller or use a collet. Steam cones are often tight in the body because of oxide build up. However, they need to be a close fit to ensure the cones align. I once was asked to test an injector where the steam cone was 0.006in smaller than the hole in the body – it was useless. Make sure you do not drop the steam cone as they always seem to fall tip down and damage the end.

Restrictions in the delivery pipework or clacks will reduce the injector range and cause water to be spilled from the overflow.

A new injector that has been properly piped up but still spills water and steam at all pressures is defective and should be checked on another loco or test boiler. If it still does not work, it should be returned to the supplier. If the injector has worked satisfactorily and then gives trouble, cleaning should solve the problem but some debris will not dissolve and may need cleaning by other means. When injectors have been used for some time, the tip of the steam cone can become worn and the cone must be replaced.

8.8 The Linden Secret, Re-starting and Lift

Part of E.J. Linden's secret was that his injectors were accurately made but the real secret, which is still not understood by many, is a radiused entry to the second part of the combining cone (the forcing tube). This gives better starting and lift than the slight chamfer that is often specified. Fred LaRoche found the secret by cutting a hole in a Linden injector to show the cones. I think that without this radius the fluid tends to bounce back from the edge of the forcing tube. Jim Ewins also used this radius.

Some people maintain that lift is not needed in injectors, however injectors that lift should restart automatically if the water supply is disturbed. Lift also allows the injector to start if it is hot, perhaps due to a leaking clack or steam valve. On miniature locos the water head at the injector inlet is low and lift helps draw water from the tank, so assists in starting.

For lifting injectors, the detailed design of the cones is important, especially the radius on the entry to the forcing tube (discussed above); the overflow valve and passage must also be of sufficient size. For a 25oz injector a $\frac{3}{16}$ in ball on a $\frac{5}{32}$ in seat with $\frac{1}{16}$ in lift is needed. The overflow passage and pipe must also be $\frac{5}{32}$ in diameter or an equivalent area. I have carried out experiments which show that restricting any part of the overflow valve or passage, or fitting a longer restricted overflow pipe reduces the lift. Most commercial injectors have smaller ball valves or passages which restrict lift even if the cones are properly proportioned and profiled. I altered a commercial injector by drilling an overflow hole through the ball cap and this significantly improved its lift.

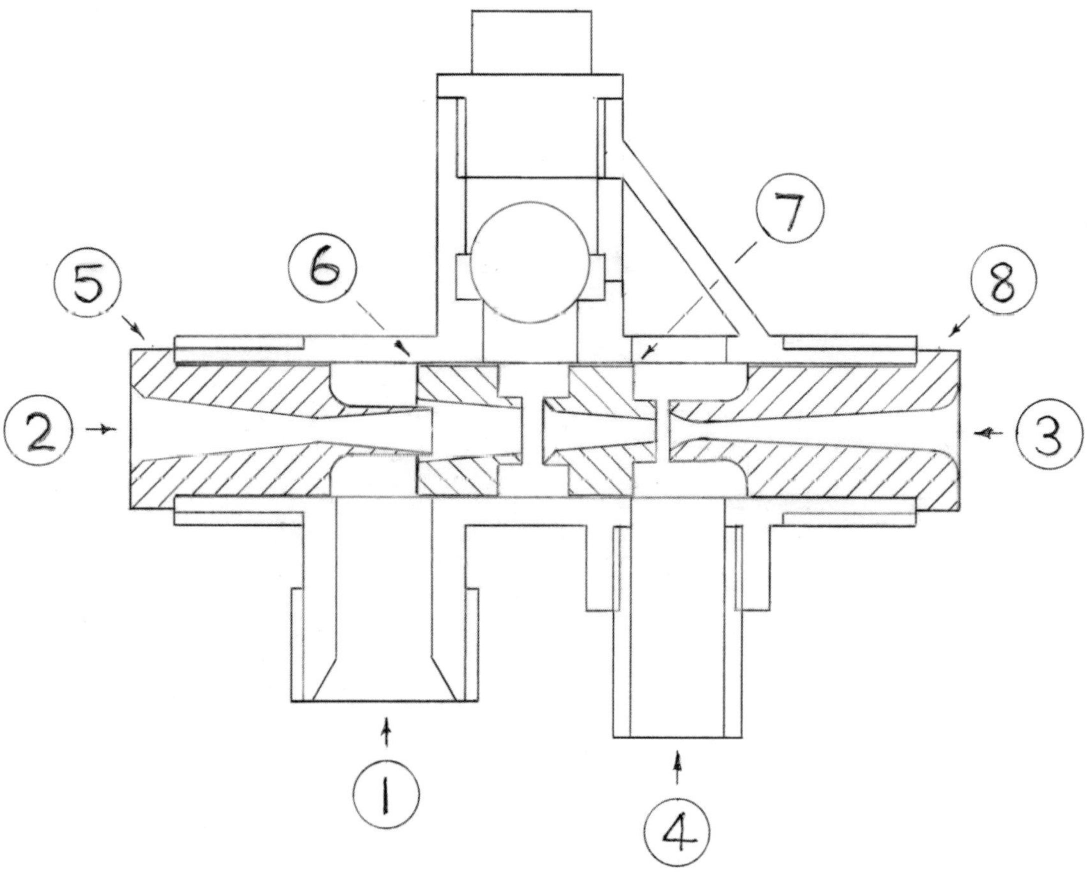

Key to Section of Injector

1	Water inlet	5	Steam cone
2	Steam inlet	6	Combining cone draft tube
3	Feed to boiler	7	Combining cone forcing tube
4	Overflow	8	Delivery cone

Cross- section drawing of a miniature injector by Richard Linkins.

See Bibliography for references.

Chapter 9

BOB UNDERWOOD

Over the years, I have been very fortunate to meet so many very ingenious and creative model engineers who manage to develop methods of solving those awkward and tricky problems in new ways. One such engineer is Bob Underwood, who modestly describes himself as 'not an engineer' even though he started building locos aged 12! Bob was fortunate to have an inspirational metal work teacher at school who ran the 'Live Steam Club'. In this after-school club, a group of lads built and ran a 3½in gauge *Iris* and Bob describes this as igniting his creative streak. He also grew up travelling to school behind Prairies and Castles and this no doubt stimulated his lifelong passion for all things GWR.

Over to Bob ----

9.1 Introduction

I love the challenge of making things and get a real buzz from driving locomotives that I have made from scratch. These will never win IMLEC or be seen in exhibitions, but they are good and reliable runners which contain some innovative ideas of my own. Whilst still at school, my first solo effort was to build a 3½in gauge *Firefly*. Once I commenced full-time employment, I changed to 5in gauge and built a 28xx, then a Saint and more recently a King.

The components described below are ones I have tried on my own engines and whose design I thought I could both simplify and improve.

Bob Underwood.

9.2 Fixing Firehole Doors

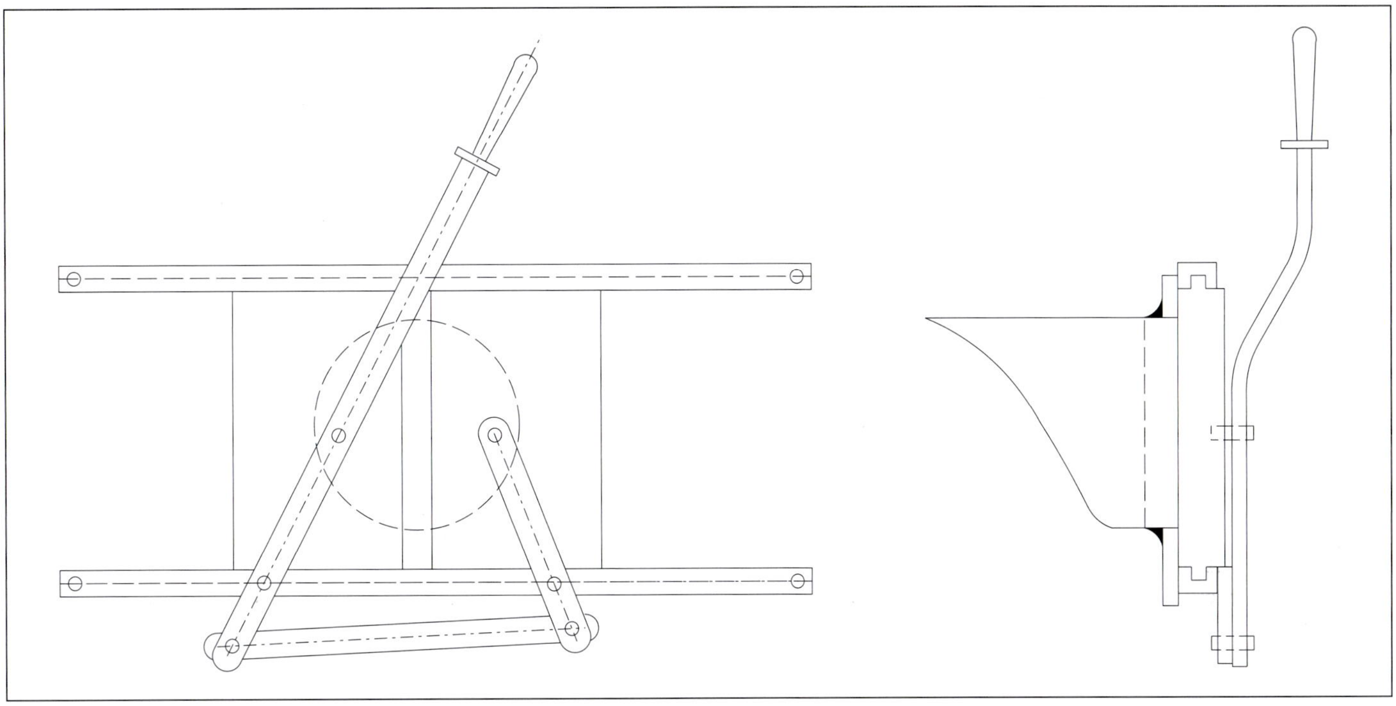

Drawing of the firehole door with shield fitted.

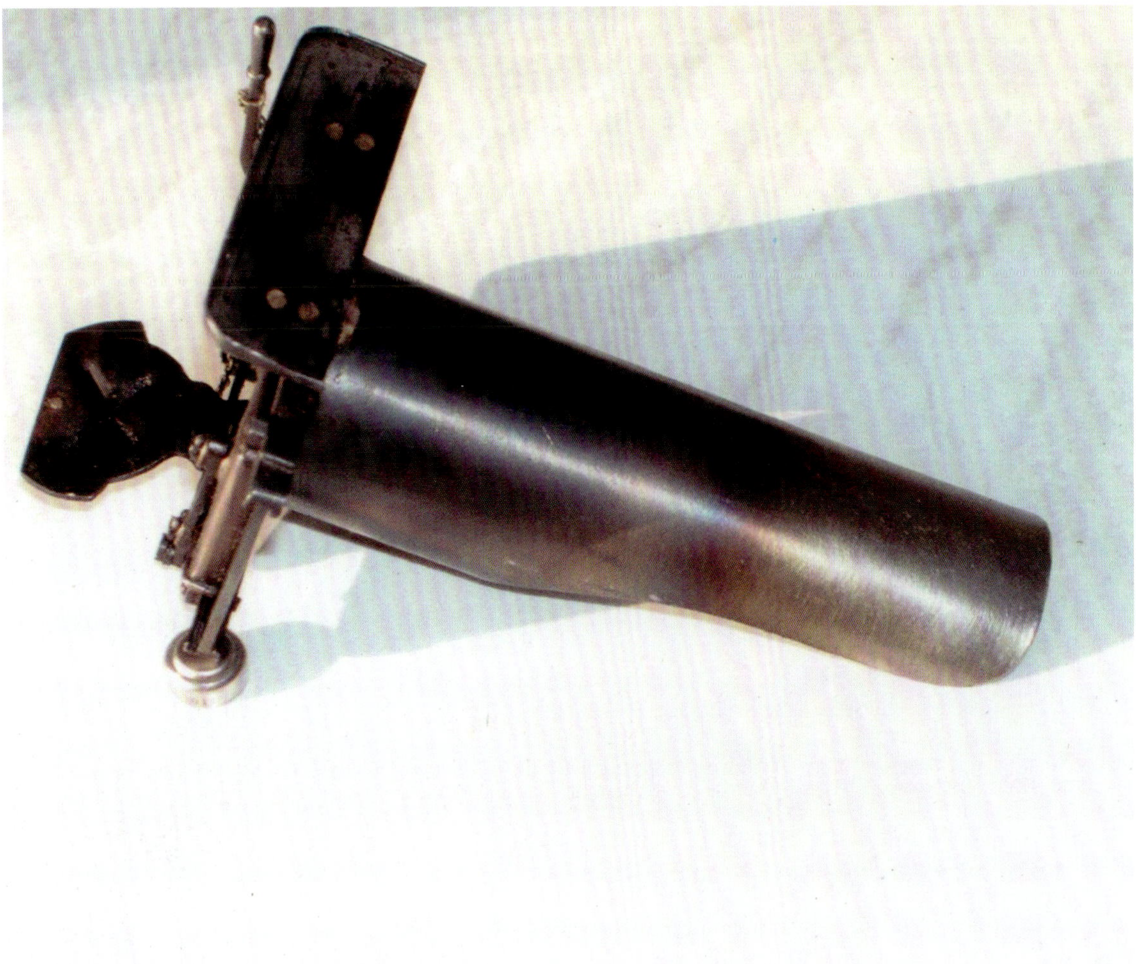

Photo of completed fire door with shield.

9.2.1 The Assembled Unit

The Firefly design called for a hinged firehole door attached by drilling into the backhead. This was clearly undesirable. I rolled a steel tube to be a snug fit in the firehole ring and the fire door backplate was silver soldered to this. For this first version, I folded the horizontal edges of the back plate to form the top and bottom door guides, but in subsequent versions I used the correct pattern of runners which were bolted to the backplate. The rolled tube also became extended into the firebox to form the deflector or baffle plate. My last engine additionally features hollow firehole doors which preheat the secondary air before entering the firebox as well as keeping the doors cooler.

9.3 Forming Seats on Ball Valves

Whilst a common procedure for forming ball seats in check valves, safety valves etc is to drill and ream the fluid bore followed by a D bit to form the base of the chamber, I have found that a better method is to make the seat independently of the valve body. This allows the hole to be drilled and reamed in the lathe and then faced off. Finally, placing a ball on paper, the fitting is simply rolled around the hole using 'spit' as a lubricant. With a little pressure a highly polished ring will appear around the seat and the reamed hole. This method has proved very reliable in practice.

Firehole doors closed.

9.4 Lubrication

Bearings on hard-working engines require sufficient lubrication to avoid premature wear, but some engines' moving parts can be difficult to access without squirting oil everywhere. All oil reservoirs were drilled as large and deep as possible so that an oil nipple just disguised their size. Oil boxes on running boards were also given as much capacity as possible with oil pipes then leading away to the bearings. Experiments followed to regulate the flow of the oil to a bearing, and it was discovered that by using a trimming made from a lamp wick, together with 5amp fuse wire, a steady flow of oil was achieved which lasted for two hours. I then modified my oil can with a $\frac{1}{16}$ in diameter outlet pipe which fitted into the same size hole in the oil nipples on the engine. Just beneath this I fitted an LED bulb which just illuminated the nipple without dazzling the driver.

Oil feed within a coupling rod boss.

9.5 Fabricated Cylinders

I decided to experiment with fabricated cylinders for my Saint class locomotive and these comprised brass plates which retained cast iron piston and piston valve liners. This arrangement has worked very successfully now for several years and avoids the cost and machining of castings.

The completed and assembled cylinder blocks.

The cross-bore stiffening webs were machined, and the liners cut to double length and passed through the webs. The double length assembly was then erected and bored out on the lathe between centres.

9.6 Cutting Multi Start Threads

Producing the correct operating system for a GWR screw reverser requires cutting a two-start left hand square thread. The scale diameter of this thread is ¼in and the cutting tool to produce a thread of this size is very difficult to make. This is because it is very small to grind, and the fast pitch requires an excessive rake angle for clearance, and this makes the tool very weak and easily broken. Having broken several such tools with numerous failed attempts, I decided to explore how this thread could be produced using the milling machine. I first screwcut a 'master' left hand thread of pitch 200 thou on a 1in diameter bar. A jig then held this bar horizontally and a pin, attached to the

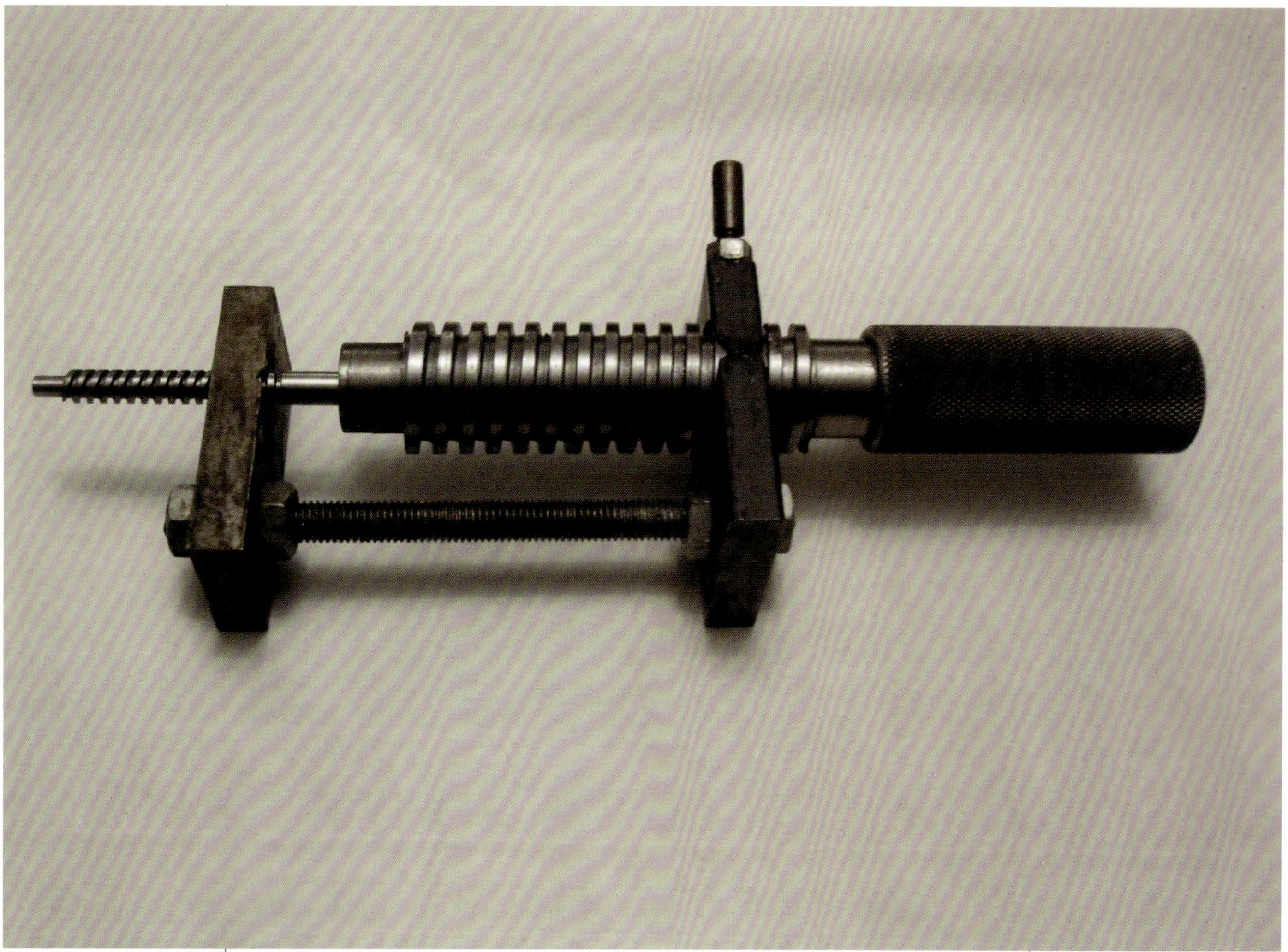

Jig for machining small multi-start threads by copying a larger diameter.

jig, caused the 1in bar with the ¼in rod to advance horizontally along the milling table bed when rotated. Using a 50 thou diameter cutter, I rotated the 1in diameter bar under the cutter and this began to mill out the first start of the left-hand square thread. Very light cuts and support beneath the cutter prevented any deflections and eventually an acceptable first start thread was produced. By advancing the milling table horizontally 100 thou and repeating the process, the second start of the thread was generated. Finally, a tap was produced by the same method which then was used to create the thread in the nut. Subsequently a friend suggested that the nut thread could be made by machining an external thread, silver soldering it into a block for the nut and simply boring out the middle!

9.7 A Removable Smokebox

Connecting pipework up in the smokebox can often be a difficult task, especially if the smokebox is long. I find the idea of cutting the smokebox on the horizontal centre line to achieve access, whilst practical, can be unsightly, so I began to consider how else access could be achieved. I explored the idea of making a fabricated saddle.

The section of the smokebox that sits within the saddle is cut out from the smokebox barrel and inserted permanently into the saddle upper curve using a spacer. This effectively forms a groove into which the remainder of the smokebox can slide and locate into its correct position. A groove is also formed across the front of the saddle

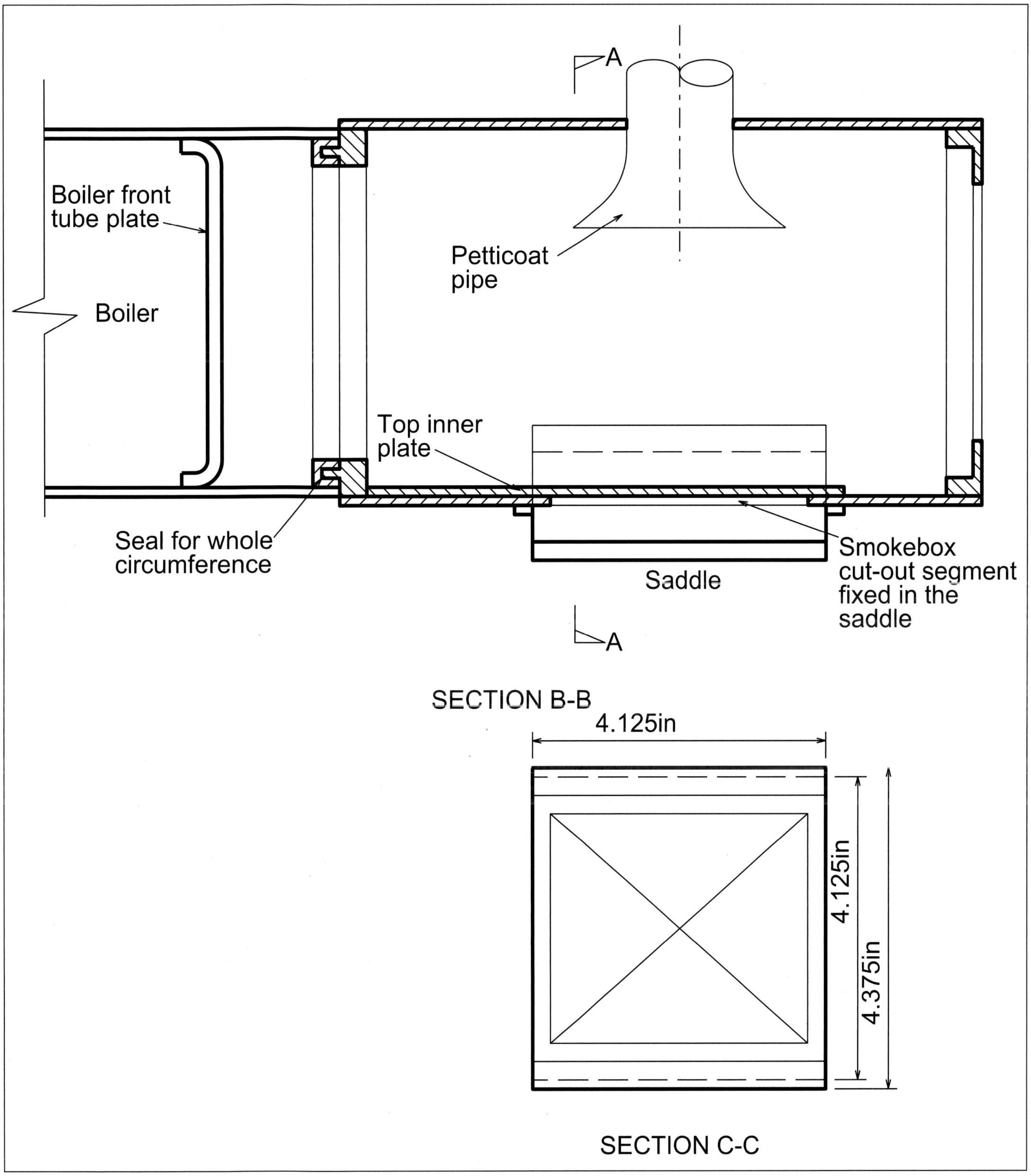

Side view of sliding smokebox.

to form a seal and similarly, a groove is formed at the junction of the smokebox and boiler such that each cut surface of the smokebox becomes slotted into a groove and hence is sealed in its final and original position.

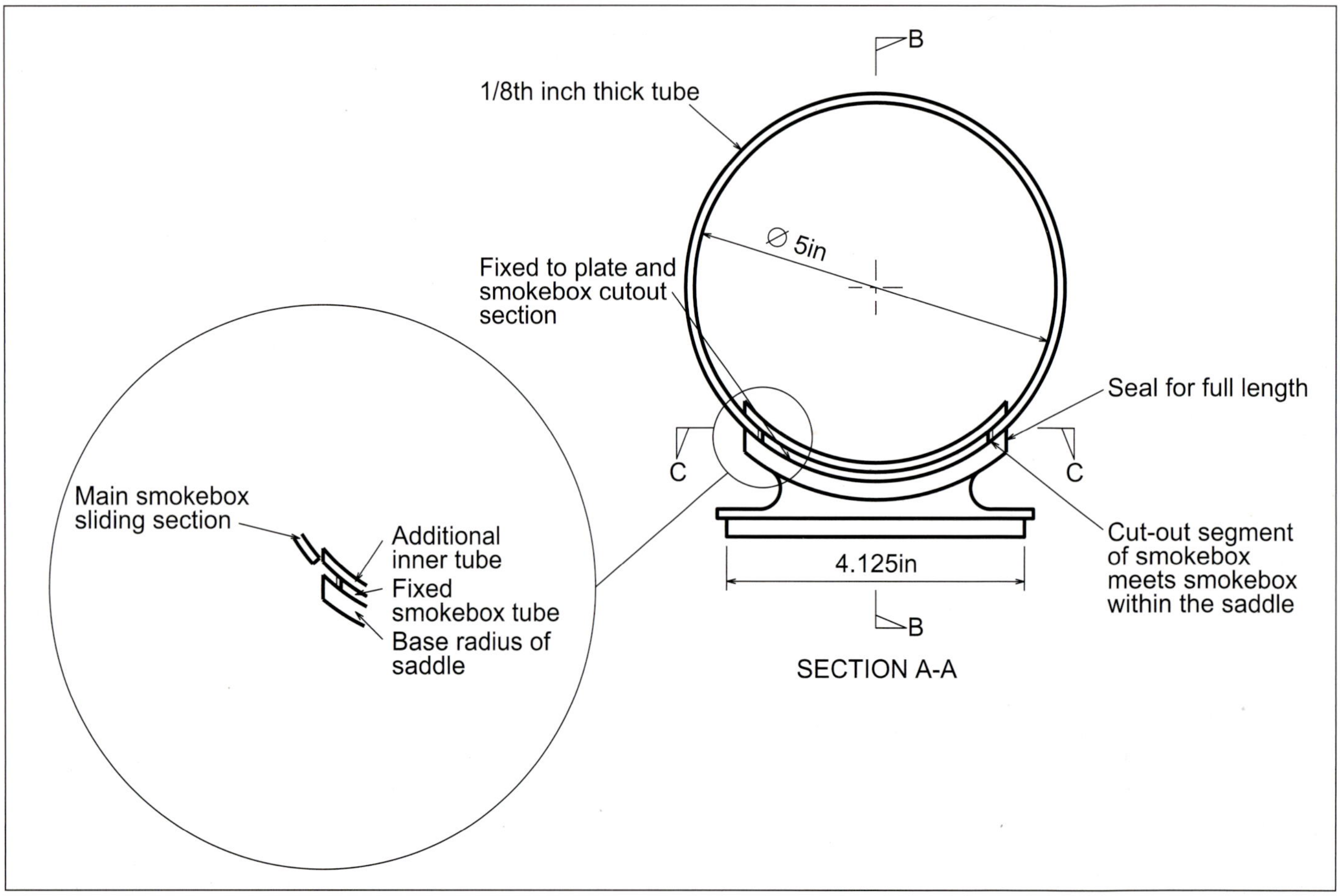

Above: Cross section of sliding smokebox.

Right: Front view of smokebox with top segment removed.

Below: Smokebox view with top segment removed.

9.8 Tender Mechanical Water Pump

I had the idea to fit a mechanical water pump to the tender of my King locomotive, delivering the feed through the same high-pressure hose as used for the hand pump. I was concerned that the high resistance of the delivery stroke might cause the driving axle to lock up and skid. I wanted the pump to be able to maintain the boiler water level at half glass when working, so I measured the water consumption of my Saint and discovered that it required a gallon of water to travel just over two miles. Making allowances for the higher boiler capacity of the King as well as the small tender wheel diameter, I built a double acting pump with a non-return valve within the piston head. On the backward stroke, water is drawn into the bore and on the forward stroke this water passes through the non-return valve in the piston head and begins to fill the piston rod side of the bore. When full, the remainder (approx. half) is displaced into the delivery outlet. On the next backward stroke, the water trapped on the piston rod side is discharged to the outlet whilst the bore ahead of the piston is recharged. Thus the cycle repeats but discharges half of the bore capacity at each of the two outputs per revolution. This has resulted in a small, high speed, efficient and reliable water pump that offers low resistance to the tender wheel drive. Seals are made using rolling 'O' rings and the drive is taken for the furthest axle to minimise angularity which would cause wear.

Thank you, Bob.
Inspired by Bob's ideas, I thought I would include a few ideas of my own that I have found useful on my own engines.

9.9 Modified Regulator

Roy Amsbury wrote a brilliant series on how to make small boiler fittings. This series is highly recommended for those keen to make fittings as close to scale as possible. My version of his regulator is just a simplification of his design, which I understand he too simplified later. The two start thread produces a valve of great sensitivity and a pleasure to operate.

Bottom: Modified regulator for the *Manor*.

Below: Finished regulator valve spindle and thread with nut above, chucking piece is still in place.

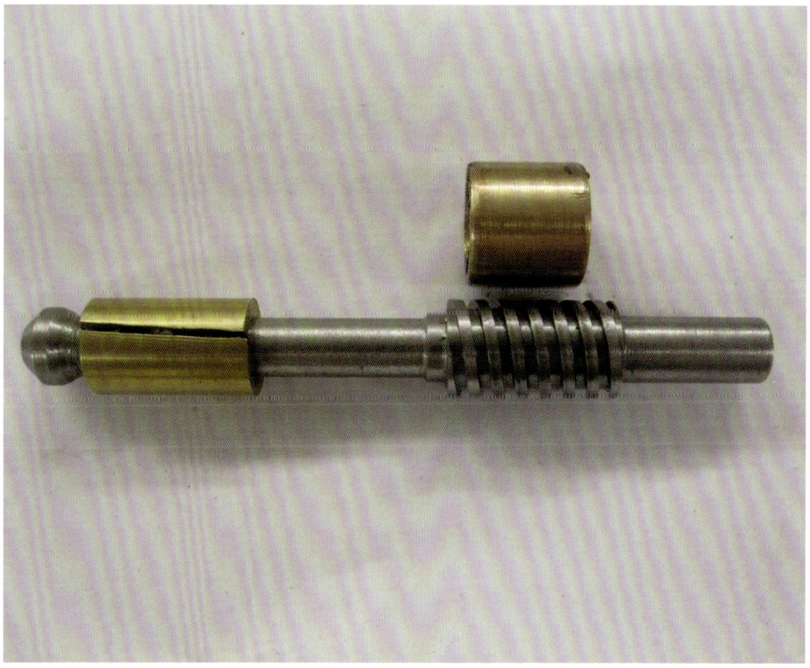

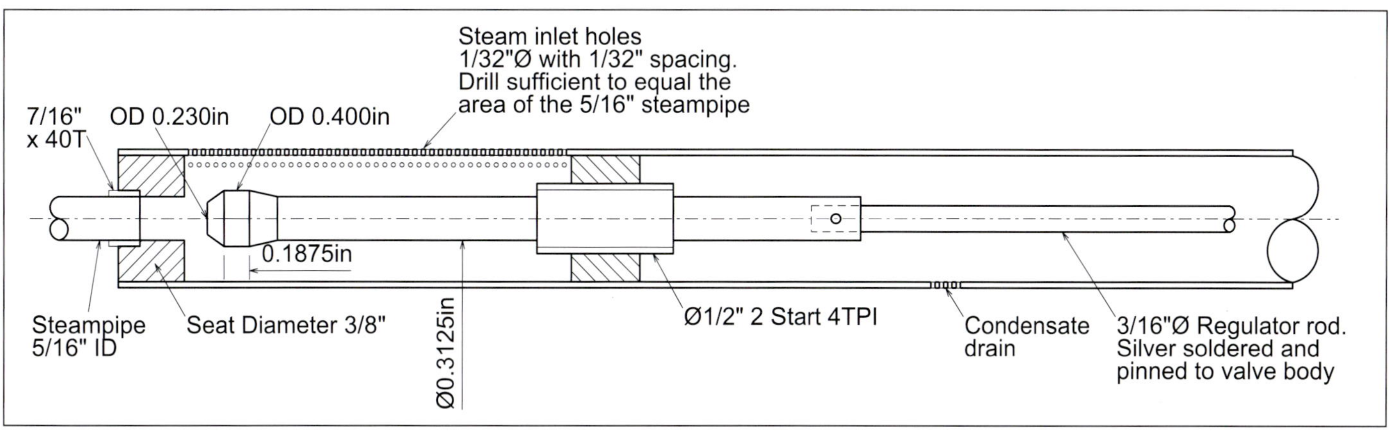

9.10 Steam Injector Valve

These follow Roy's designs closely but feature a threaded connection into the boiler instead of the flange that Roy shows. Otherwise, his design has worked well for many years.

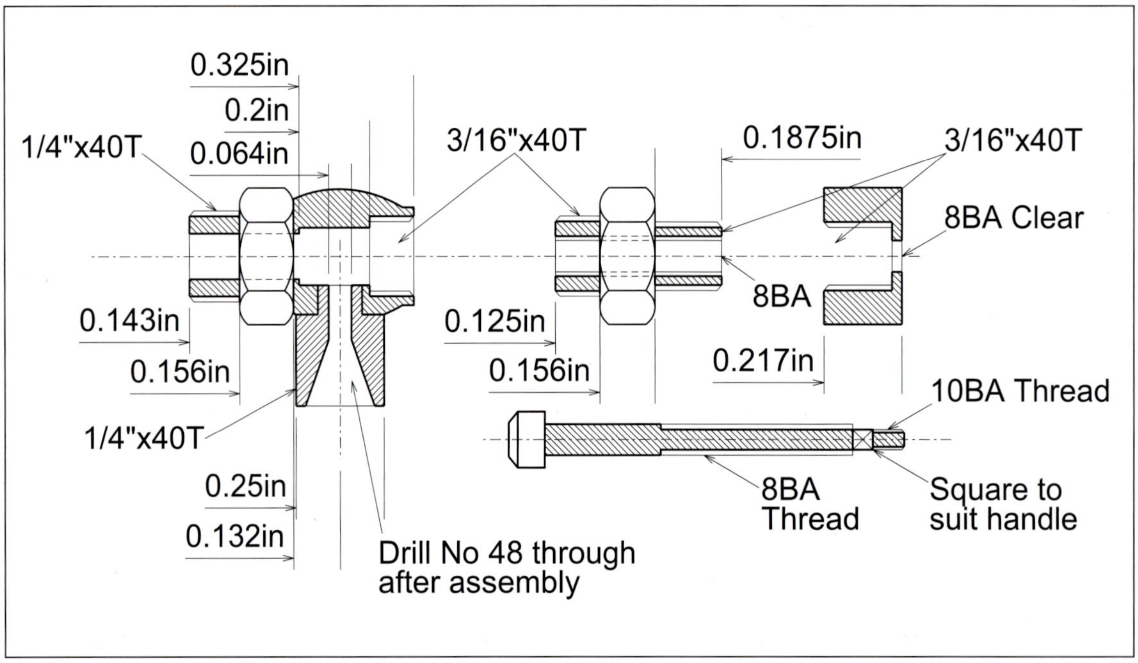

Above: Component parts of Roy Ambury's modified injector steam valve.

Below: Assembled parts of Roy's modified injector steam valve.

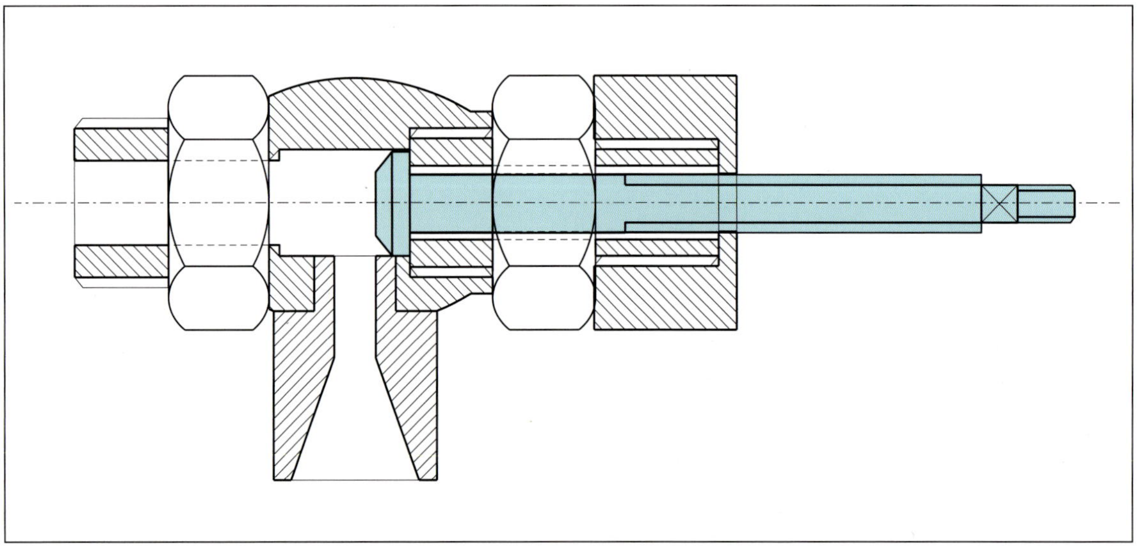

9.11 Safety Valve/Top Feed Assembly

Below is copy of Roy's drawing for this important and quite complicated fitting as well as a photo of my own version which follows Roy's design except in two areas.

I prefer ball seatings to be removable so the ball seats for the clack valves are contained within a basket which is removable from the top. I have never needed to remove this to service the seats, but it is a useful maintenance facility.

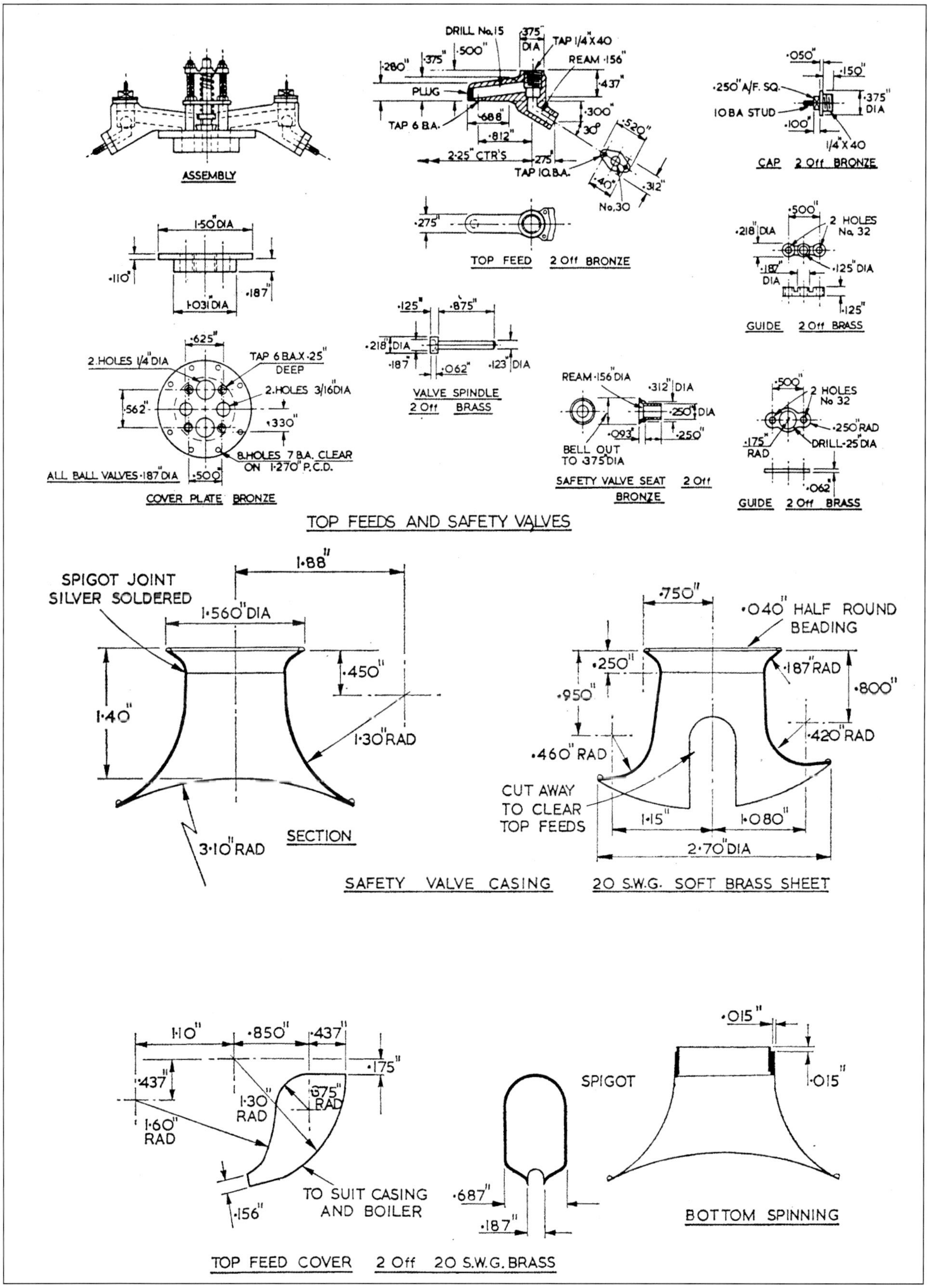

Roy Amsbury's drawing for the top feed and safety valve.

Modified safety valve assembly.

I had heard of instances where the safety valves had lifted and caused water from the injectors to be picked up and discharged straight out of the safety valves. To mitigate this, I screwed small elbows into the underside of the fitting and then installed delivery pipes that carried the injector feedwater some 2in towards the front of the boiler. The system has never given any issues.

9.12 Steam Manifold

It is often tricky to ensure that minor valve seats, like the whistle valves, can pass the high hydraulic test pressures without dribbling. To mitigate this, I always fit the rear of the manifold with an elbow that has an internal valve seat, which is accessed from the top leg of the elbow. By removing the top blanking cap and screwing in the valve stem shown, the manifold can be completely isolated, so removing the need to even remove the pressure gauge when hydraulic testing. The elbow is completely hidden behind the manifold so there is no detraction from the prototype appearance.

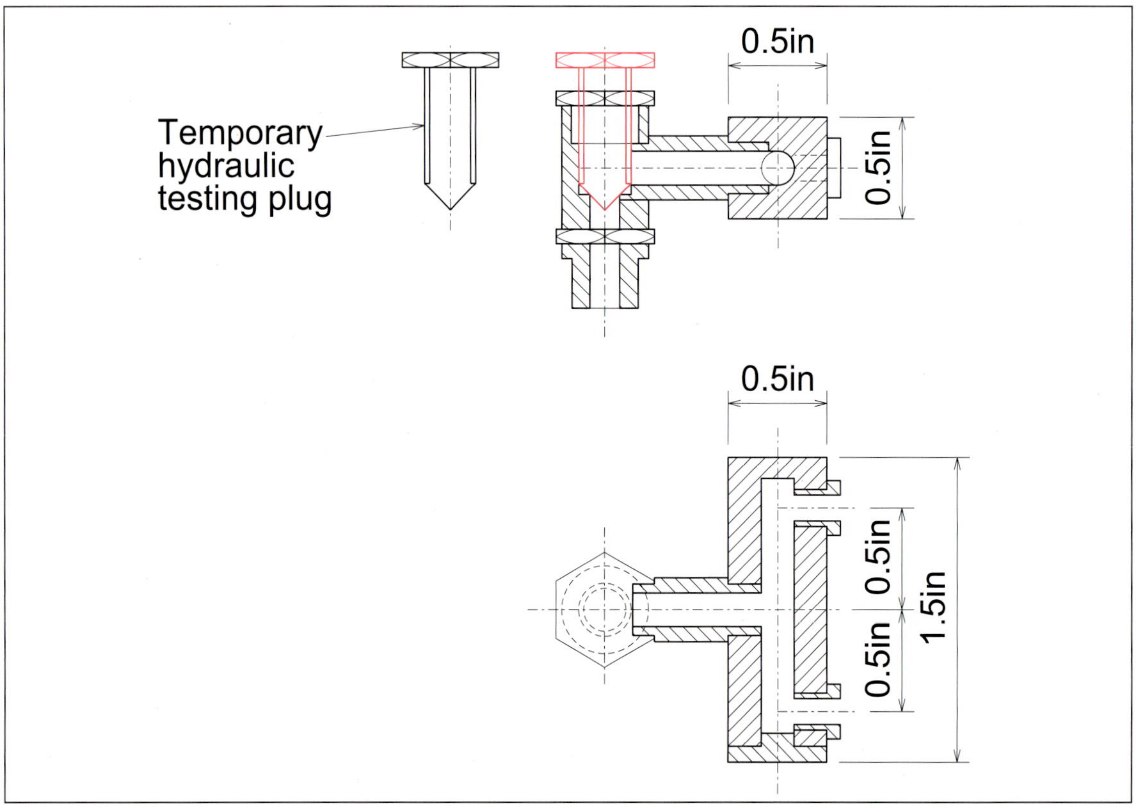

Assembly drawing of the manifold and isolating valve.

Side view of the manifold.

Front view of the manifold in steam.

CHAPTER 10

BATTERY OPERATED LOCOMOTIVES

10.1 Introduction

With the increasing popularity of miniature passenger hauling railways, there is a demand to be able to have an engine that can be brought into use and disposed of quickly. Several companies now offer 'ready to run' battery powered engines in many outline guises, gauges, and power outputs. The ability to have an engine 'ready to go' can be very useful, even on 'steam only' lines. They can be used for route knowledge, driver training, shunting and works trains. They also allow someone without workshop facilities or skills to enjoy running trains on their own railway.

Ben Harris driving the Class 47 on the new Welling Club track at Hall Place, Bexley, Kent.

For all these reasons, my own railway now has two battery locos on shed, with a third, (a Class 66) under construction. Many model engineering club tracks also have several battery powered locos for the same reasons. Many younger enthusiasts also prefer to see modern traction locos as it is the railway scene they have grown up with.

When my own garden railway was under construction, I realised that the 08 would be a useful loco to have available. This design has proven to be reliable and powerful. It now moves materials around the railway as well as giving tours for visitors. A second electric locomotive was purchased from the Great Wealden Railway when it was dismantled, and this had the additional feature of being radio controlled. Initially, I thought that this completely missed the pleasure of driving a train, but it does have its uses and has been fun and enjoyed by junior drivers.

The following are some notes on building an electric locomotive from scratch.

Two motor drive for the Charlatan 08 Class diesel shunter.

10.2 The Chassis

This can be a similar construction as for the steam loco chassis, depending on the type of loco being built. This approach was initially used in full size by many industrial builders as well as BR, and the 08 is a typical example of this.

The location of the electric motor(s), as well as the space consuming batteries, will dictate where frame stretchers need to be placed. The 08 has motors mounted on its axles by use of arms that are supported by bearings on the axles. A third arm connects to the chassis to absorb to the motor's torque reaction. The engine's driving wheel is driven by a gearwheel fixed directly to the back of the wheel and this arrangement has the following advantages:

- The motor drive remains in constant connection with the wheel set and is independent of the movement of the suspension.
- The low-slung motors keep the centre of gravity of the loco as low as possible.

Single motor drive for a Class 66 axle.

Some commercial designs feature a cast aluminium chassis which is best suited to mounting motors on top of it with a chain or belt drive to the axles.

Phoenix Class 14 with motor on chassis and belt drive to an axle.

The following observations are provided largely by the company 4QD, who supply excellent control systems. All my locos incorporate them, and they have been trouble-free. Their website is full of useful information, and they are always helpful.

10.3 The Electric Motors

The size and number of electric motors needed for an engine is determined by the required tractive effort and speed. Also consider the acceleration required and the gradients on the route. The Charlatan design uses four No. 12V Lucas 78550 motors and these have proven to produce a very powerful locomotive with all the speed required for a 5in gauge engine. Should the load demand a current more than 30-50 amps, it is recommended that 24V motors be used to keep heat generation and losses to a minimum.

10.4 The Control System

Always match the control system with the motors to ensure complete compatibility.

Earlier controllers were designed to use relays, but arcing across contacts is known to shorten their operating life. 4QD now mainly use MOSFETs in the control circuits and these have proven very reliable, stable, and fast in action. Their units also provide adjustments for ramping up and down, a braking and sound chip option. All very realistic and straightforward to wire up.

10.5 Batteries

Batteries are the source of power for electric locomotives and there are several types which appear suitable, but in fact, are not.

Car batteries depend on being constantly charged in use and this is not practical on our miniature electric locos. These batteries will run until discharged, but successive re-charging causes a build-up on the battery's internal plates which increasingly adds resistance to the cells. This progressively reduces the current flowing and ultimately these batteries will fail to re-charge and will need to be replaced. There is a range of commercial batteries that are designed for being fully discharged and they are known as 'Heavy discharge / deep cycle' batteries. One such popular version of these is the Lucas LSLC42-12 and these are recommended and used successfully by the company AME Railways. With both the car and heavy discharge batteries, regular 'trickle charging' is the best way to preserve their life. Ensure that the battery feeds are easily disconnected for the charging process and eliminate the risk of equipment receiving a voltage it should not.

Always talk to electric locomotive suppliers who will ensure that your locomotive provides the maximum hours of trouble-free running.

Amelia enjoying a drive along the line.

Edward keeps an eye on Chloe as sister Nora looks out.

The remote controlled Class14 on a freight train.

Hugh discussing the remote control with Thomas.

Barry enjoying driving a full train on the Bentley Miniature Railway Uckfield.

CHAPTER 11

BUILDING THE RAILWAY

11.1 Making the Track

The following describes the construction of 5in gauge track, but the techniques will apply to other scales with minor modifications.

Historically, elevated tracks were built using rectangular steel for rail mounted on 'A' frames. In recent years, ground level tracks have become popular with commercial passenger lines constructed in 7¼in gauge and upwards. The GL5 Society have done much to encourage 5in gauge ground level layouts and these feature points and crossings and prototype operations.

Paul enjoying a local freight at Great Wealden Railway, Biddenden.

John Bentley preparing the Manor for the road. Michael Cadiz is behind left.

11.1.1 Types of Rail

Prototype steel and aluminium rail sections are available, and the choice depends on usage and cost. Aluminium has been excellent for my own ground level railway; it's easy to work, looks good, doesn't rust, and offers sufficient adhesion for moderate passenger hauling.

11.1.2 Types of Sleepers

The cheapest form of wooden sleeper is the household roof batten. Being tanalised softwood it offers a good life and is readily available. With a cross section of 38mm it is wider than scale for 5in gauge but the Great Wealden Railway used this material, and it lasted for some fifteen years. I used Douglas Fir to a scale width for my own line and, with treatment, it has worked well.

Several companies now offer extruded plastic sleepers and PNP also supply 'plug in' chairs, making track construction simple and fast.

11.1.3 Bending, Fixing Rail and Joining Rail

Aluminium rail bends easily and the layout and site will dictate the maximum radius possible. My own railway uses 10m curves and my 4-6-0 Manor has no problems with these. I made a simple three point bending device based on an old machine vice and it has bent all the curved rails on the railway.

The distance between the two fixed rollers must be kept to a minimum as this produces a length of straight rail end of each curved section.

My railway was built using 8ft track panel lengths which were assembled on a

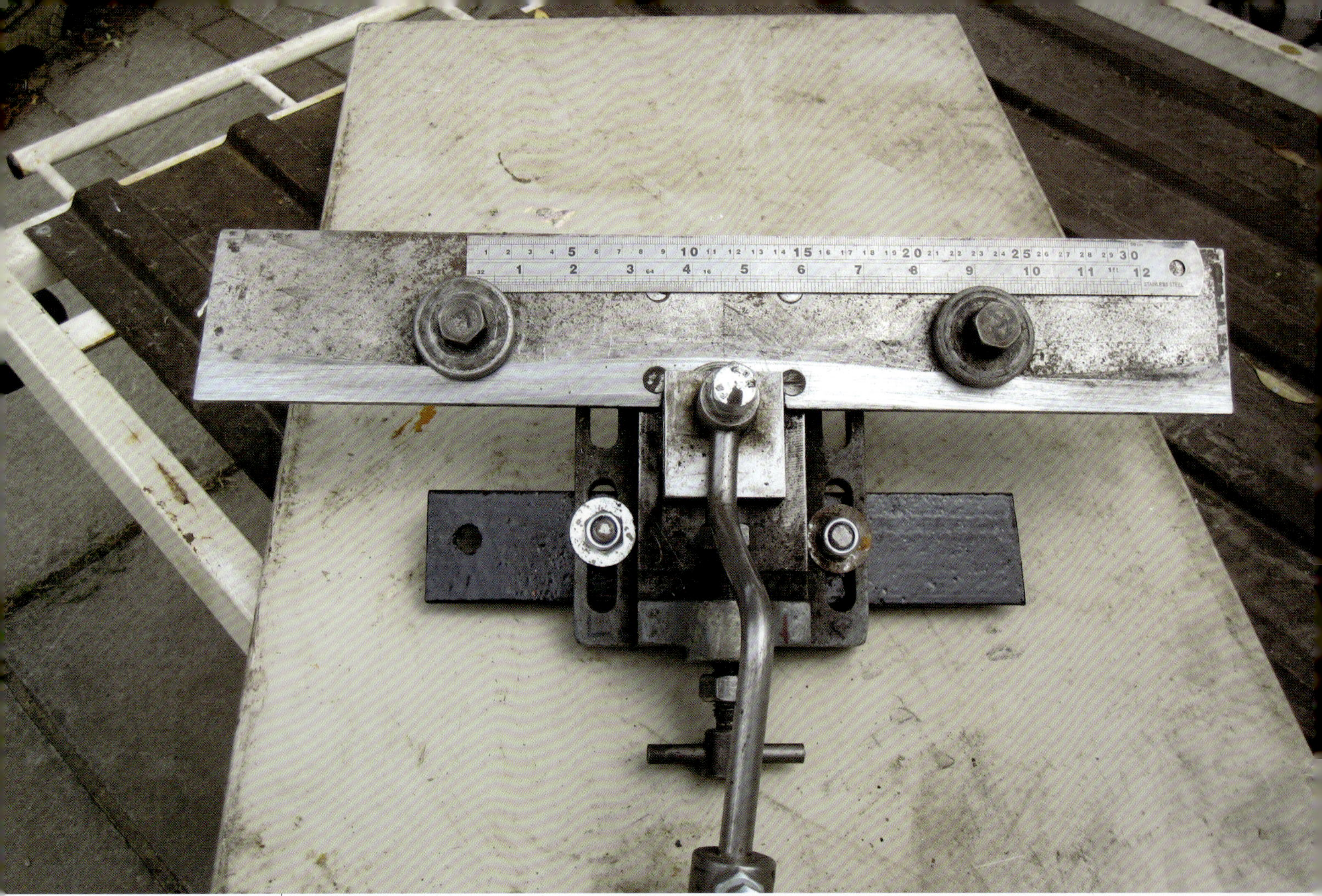

Three-point track bender.

Rail gauging and fixing in progress.

bench and then taken to site. Sleepers were spaced at 3in centres and templates used to set out all panels. Using gauges to lay out the sleepers and rails, a completed panel took 45 mins.

I used 4mm x 25mm fine stainless-steel screws for all rail fixings and M4 stainless steel Nyloc Nuts and bolts to connect the rails using stainless steel Fishplates. A CAD drawing was produced for the fishplates, and these were cut out on mass by water jetting.

11.2 Earthworks

An average 5in gauge loco with rolling stock and several passengers will exert significant forces into the ground via the track, and these must be resisted by the track formation. The type of ground along the route will determine whether or not a drainage system needs to be installed. A formation built on undrained clay, for example, will quickly become flooded with heavy rain and begin to look more like a canal if a drainage system is not installed. Chalk provides a strong structural support for the formation and will not require drainage. Shales and sandy materials will provide a much lower ground bearing pressures but will also not require drainage.

Commercial drainage systems are available in the form of large coils of perforated plastic pipe of 75mm diameter and this should be installed on the centreline of the route for the formation. The system can be run off to a sump or holding tank.

Left: Track curve section ready for installation.

Below: Standard stainless-steel fishplate used throughout the railway.

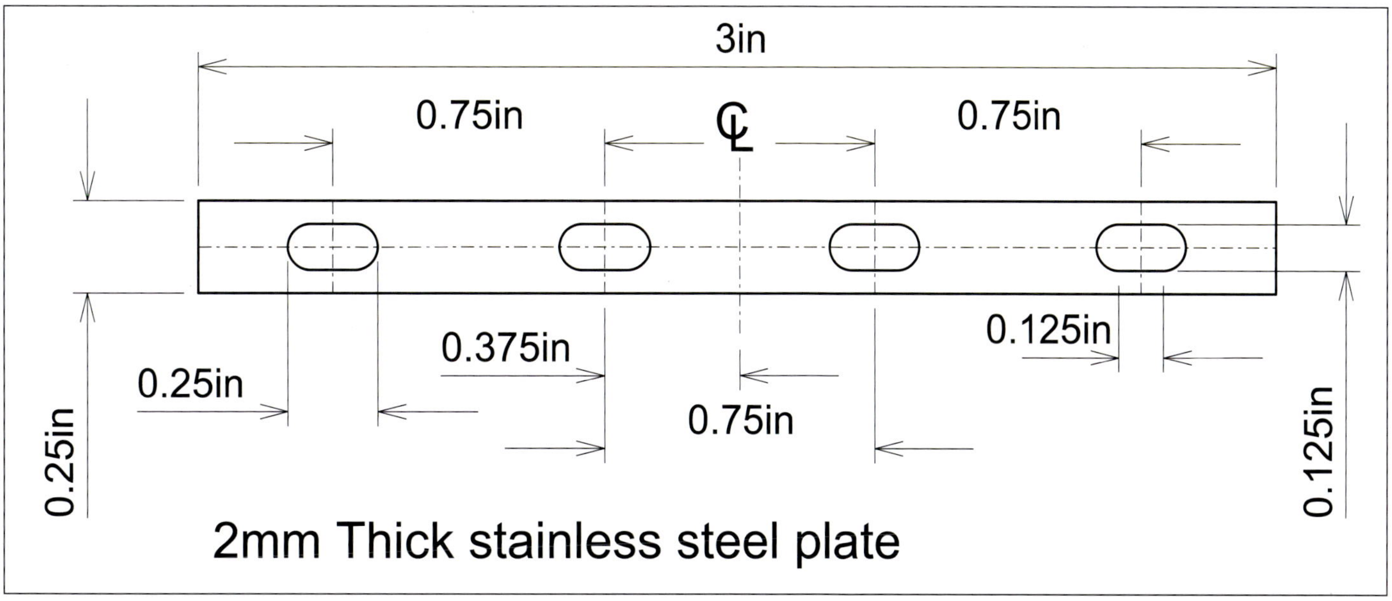

Tony and Tom checking final earthworks at the site for Welshpool Station.

Tom working on site clearance in progress at Triangle Junction.

Tony blending the Embankments.

Tony Rea digging final levels at Triangle Junction.

Tony and Tom using a vibrating plate to compact the ballast at the location for Llynclys Station.

Ballasting through the station.

Oswestry Station trackwork in progress.

Track preparation for Triangle Junction.

Derek Brunning supervises Tom ballasting track at Triangle Junction.

The maximum number of passengers on a train will determine the ruling gradients on the line and this, in turn, will determine the extent of the earthworks. One section of my own railway has a gradient of 1 in 80 and this has worked well for moderately loaded trains. Earthworks are expensive, so survey the route carefully and ensure that you have chosen a datum that creates the minimum volume of earthworks.

11.2.1 Laying out the track.

Full-size track has a real elegance and I wanted to capture this for my garden railway, which is based on the Cambrian system. Inclusion of key junctions and stations do allow similar train movements to the prototype.

11.3 Steaming Bays and Movement of Engines

For easy servicing of locomotives, elevated steaming bays were a must for me. These were built from 2x2in steel angle at waist height. Trolleys of the same section are used to move and store engines and align with the steaming bays. Trolleys also ensure safe movement of engines around the railway.

11.3.1 A turntable that elevates

I have been using hospital beds as mobile elevating benches for years and it became the obvious starting point for an elevating turntable. The width of the bed was reduced to 18 inches and then mounted on a Mini front wheel bearing to achieve rotation. Removable arrestors fit to each end to prevent engines rolling and the system has worked well.

Steaming bays.

Turntable down.

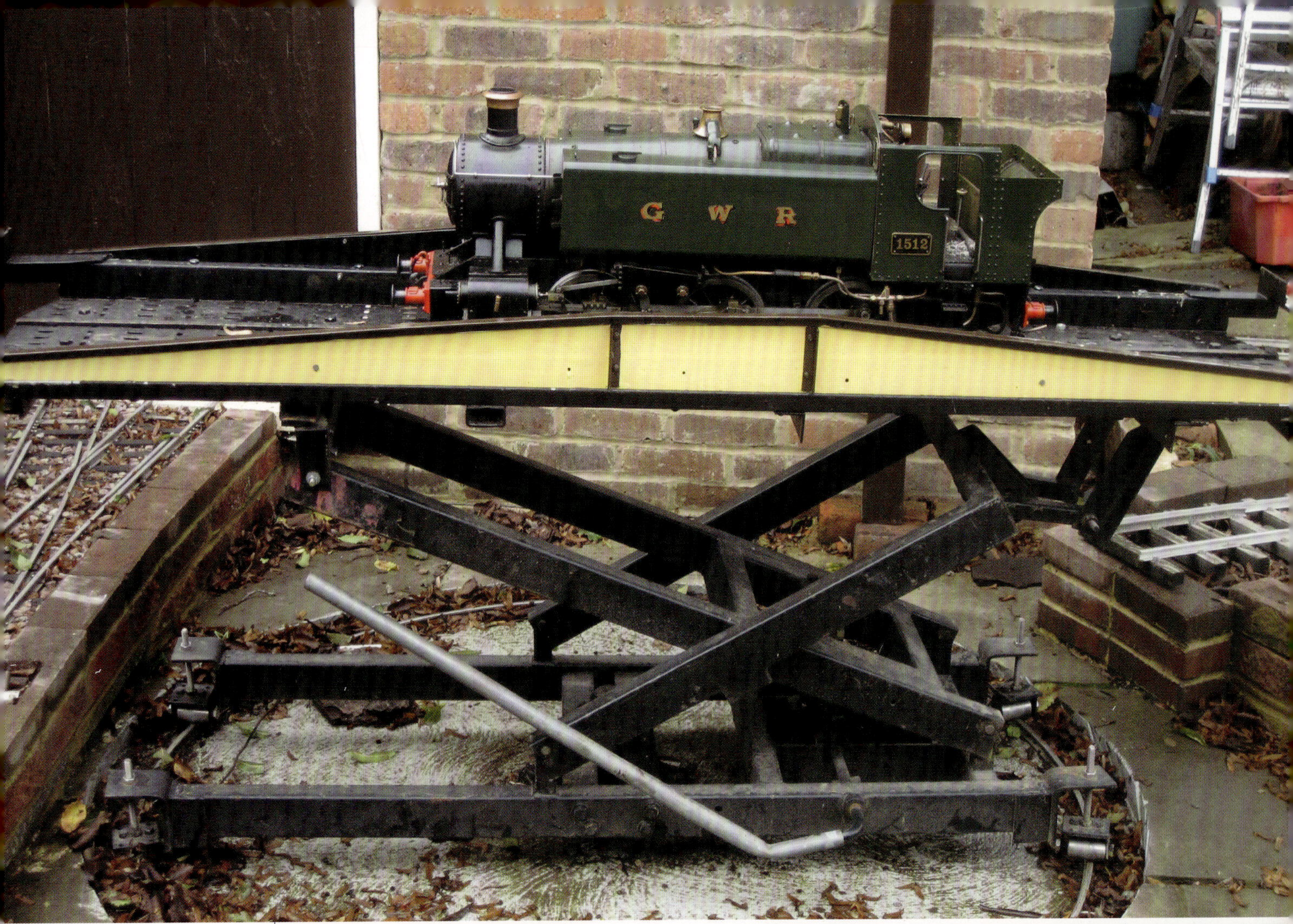

Turntable Up.

Y Point on entrance to Triangle Junction.

Hugh building a five-way junction leading to the carriage shed.

Five-way junction installed.

Double slip installed.

Upper shunting yard completed.

11.4 Making Turnouts

These are a major feature of all full-size lines and allow the railway to have passing loops, sidings etc, just like the real thing. These are straightforward to make as described below:

11.4.1 The Crossing Nose

The best way to manufacture these is to follow full size practice. The primary route rail is cranked which ensures that the nose of the crossing is formed around its web. Machine the side of the primary rail to allow the secondary rail to be fixed into it, bolt the two sections together and, using straight edges, file the rails to form a flowing V.

11.4.2 Machining the Switch Blades

The leading section of the switch blades becomes very thin when machined to fit into the stock rail and it is essential to make a jig to fully support the whole length of the machined side of the rail during the milling process.

11.4.3 Machining the Stock Rails

The stock rails also require supporting whilst milling the relief to allow the switch blade to sit flush within the stock rail. Again, a jig is essential and final fettling with a file will ensure wheels pass smoothly.

11.4.4 Assembling all the parts

Once machining is complete, build the turnout in the following sequence:

- Produce a 1:1 drawing of the finished turnout.
- Lay out all the sleepers required to form the turnout on the drawing.

Construction of crossing joint.

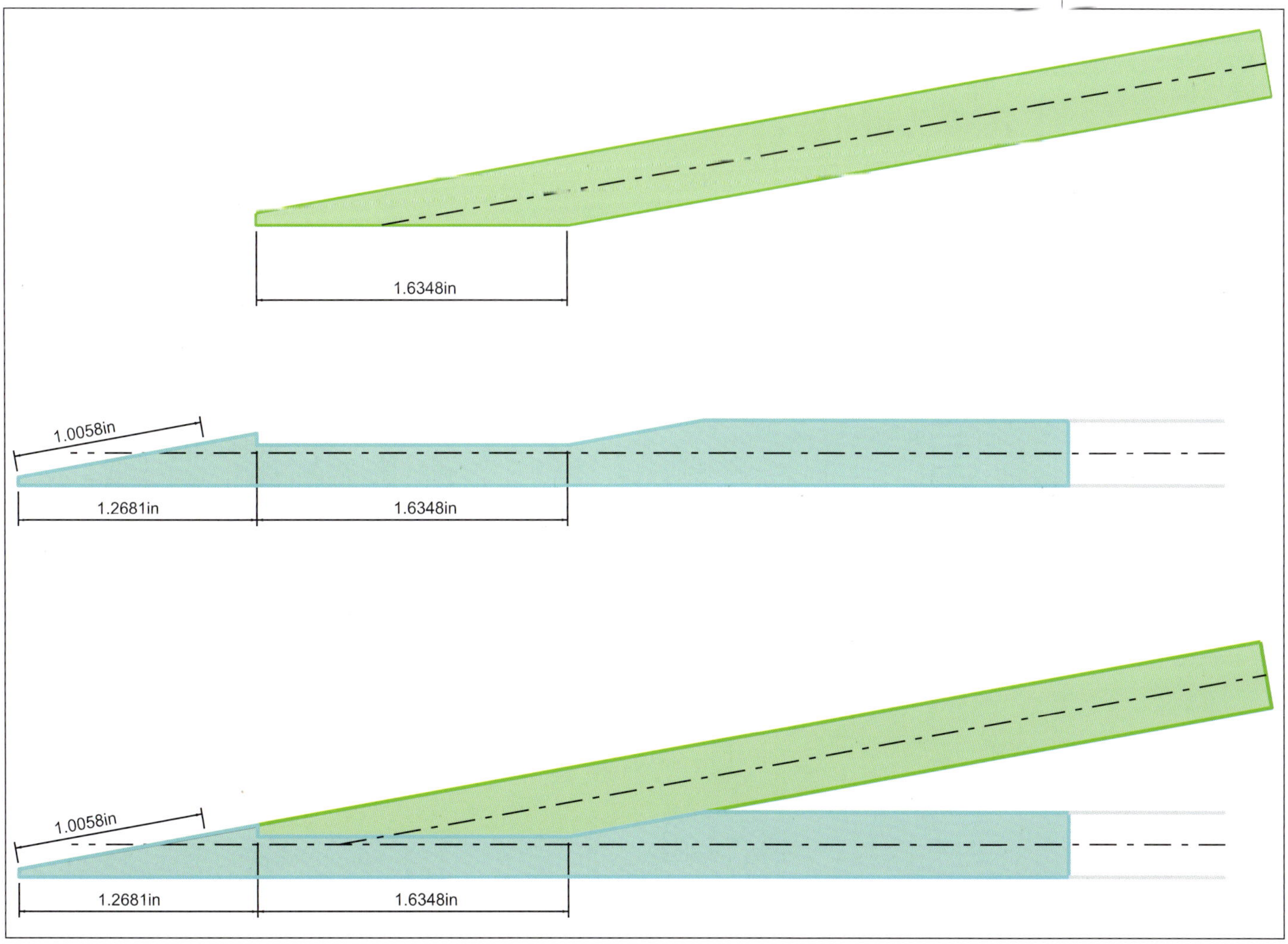

- Fix the straight stock rail in position to the sleepers.
- Temporarily place the curved switch blade against the straight stock rail to locate the position of the crossing Nose.
- Use track gauges to ensure that the straight side of the crossing nose is at 5in from the straight stock rail and parallel to it. Fix the crossing nose in position.
- Repeat the procedure to locate the curved stock rails and fix in position using track gauges. Again, use the straight switch blade as a check for position and gauge.
- Offer up the wing rail sections of the switch blades to form the crossing nose check rails and fix in position with a 7⁄32 in gap from the nose.
- Trim the lower sections of the switch blades to length to meet the joint with the upper section but do not install yet.
- Fit check rails to both stock rails and check that any long wheelbase vehicles pass without binding.
- Drill all the fixing angles between the stretcher bar and the two switch blades, noting that final assembly for these must be in-situ on the turnout.

11.5 Controlling the points

Having provided turnouts, a means of controlling these is required. A mechanical lever at the side of a turnout allows the train driver to manually select the route. Some turnouts can be sprung loaded, providing route selection and a 'trailing through' capability. By using two opposing springs on the stretcher bar with a bias selecting lever, either route can be pre-selected. Ultimately powered turnouts driven by electric motors from a signal box is the prototypical solution but requires a signalman.

Martin Evans has designed an excellent control system for his own railway which features train detection and route selection This is described in the *Model Engineer*.

See Bibliography for details.

Trailable point with adjustable bias.

CHAPTER 12

MAKING ROLLING STOCK

The pleasures of a ground level railway are maximised when trains comprising prototypical rolling stock completes and recreates 'the scene'.

12.1 Wagon Construction

Wagons comprise a basic chassis from which 'W' irons, axleboxes and wheels support wooden or metal body. Early wagons used a wooden chassis, but steel sections are more common and practical allowing strong fixings for the 'W' irons. Rod stays connect the 'W' irons, creating a light but strong space frame. Works drawings of wagons are so informative and should be used as a guide. Axlebox machining and wheel turning are the same processes as locomotives, but axleboxes are often fitted outboard of the wheelsets. My Toad guards van doubles as a driving trolley, so strengthening the body and suspension for this load needs addressing. Bogie Wagons are now common and are constructed like coaches. Doug Hewson has published many highly detailed plans for wagons, and these can be recommended and found in the leading magazines and publishers.

Excellent wagons and trucks can be purchased commercially, and the photos below show some superb examples from Aristocraft:

GWR Toad in 7¼in.

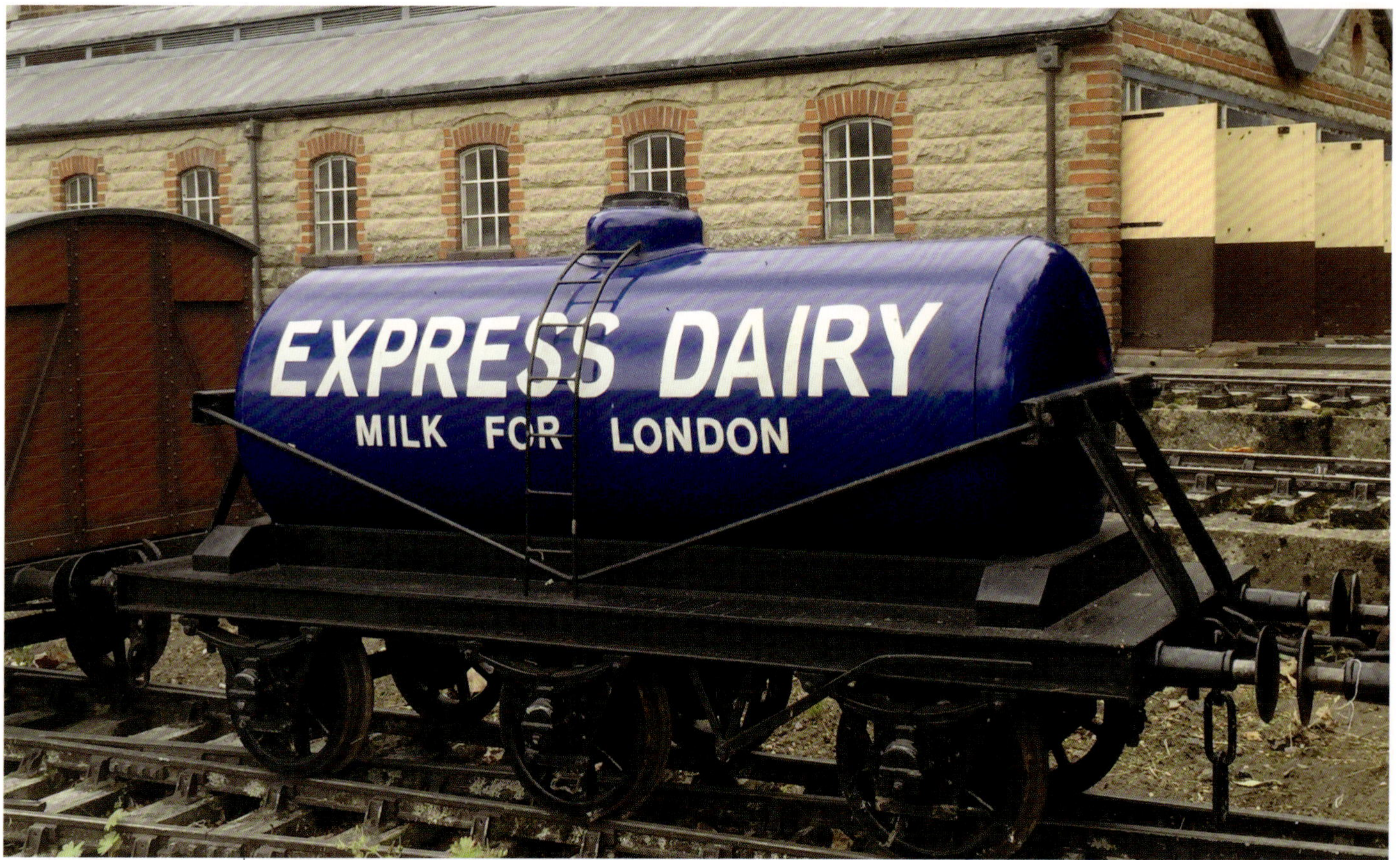

Express dairy wagon in 7¼in.

GW 12T ventilated van in 7¼in.

Side view of Toad.

12.2 Coach Construction

Coaches are large structures which require strengthening to carry passengers. Coach construction can be described in three basic sections, the chassis, the superstructure, and the bogies.

12.2.1 The Chassis

In 5in gauge, a typical coach is in the order of six feet long, which could accommodate three adults, so the chassis must support this load without deflection. 1x1in steel angle is good for a chassis as is 1in square box section. The battery box racking/lattice framework beneath the chassis performs stiffening by adding depth at the weakest central point. Some miniature coach bodies have been made into an integrated box construction with the roof, sides and floor forming a one-piece assembly. Whilst strong and not requiring a chassis, it makes access into the interior difficult, especially for glazing the windows.

I build a separate chassis for my coaches, and this allows body and chassis to be separated for construction, painting, and maintenance. The lattice underframe can be made from ½x½in steel angle and the whole chassis welded, or 'Sif-bronze' welded together.

12.2.2 The Superstructure

The coaches I chose to model were standard BR Mark 1 coaches. Side panels of the coaches were laser cut from 2mm steel sheets, together with window frames. The coach sides have a large radius from top to bottom and this was rolled by a local fabricator. Doug Hewson has written a detailed series on building Mark 1 coaches in the *Model Engineer* magazine and provides useful information. See Bibliography for details.

Mark 1 Coach body in steel by Paul.

Rake of BR Mark 1 coaches in blood and custard livery.

Aristocraft vintage GWR coaches on shed.

Rake of BR(W) Region coaches.

Rake of homemade Mk 1 coaches with Hugh driving the 08.

Some builders create fibre glass superstructures for their rolling stock and such moulds are excellent if several vehicles of the same type are to be constructed. Several commercial companies also sell this type of body.

12.2.3 Bogie design

Initially my bogie design followed the tradition of four independently sprung axleboxes and conventional mainframes. This has given good service but involved a lot of machining. Aristocraft sell a sit-on coach and its ride quality impressed me. Instead of using axleboxes, it features the use of spherical ball races mounted on steel mainframes. These are pivoted at their centre by a circular cross beam which can rotate fore and aft. The bogie centre pivot passes through the cross beam, so, together with outboard stabilisers, this arrangement accommodates movement of all the wheelsets in all three planes.

12.2.4 The Buffing System

It is desirable that coaches are closely connected to create the look of a close coupled full-size rake, but it is essential

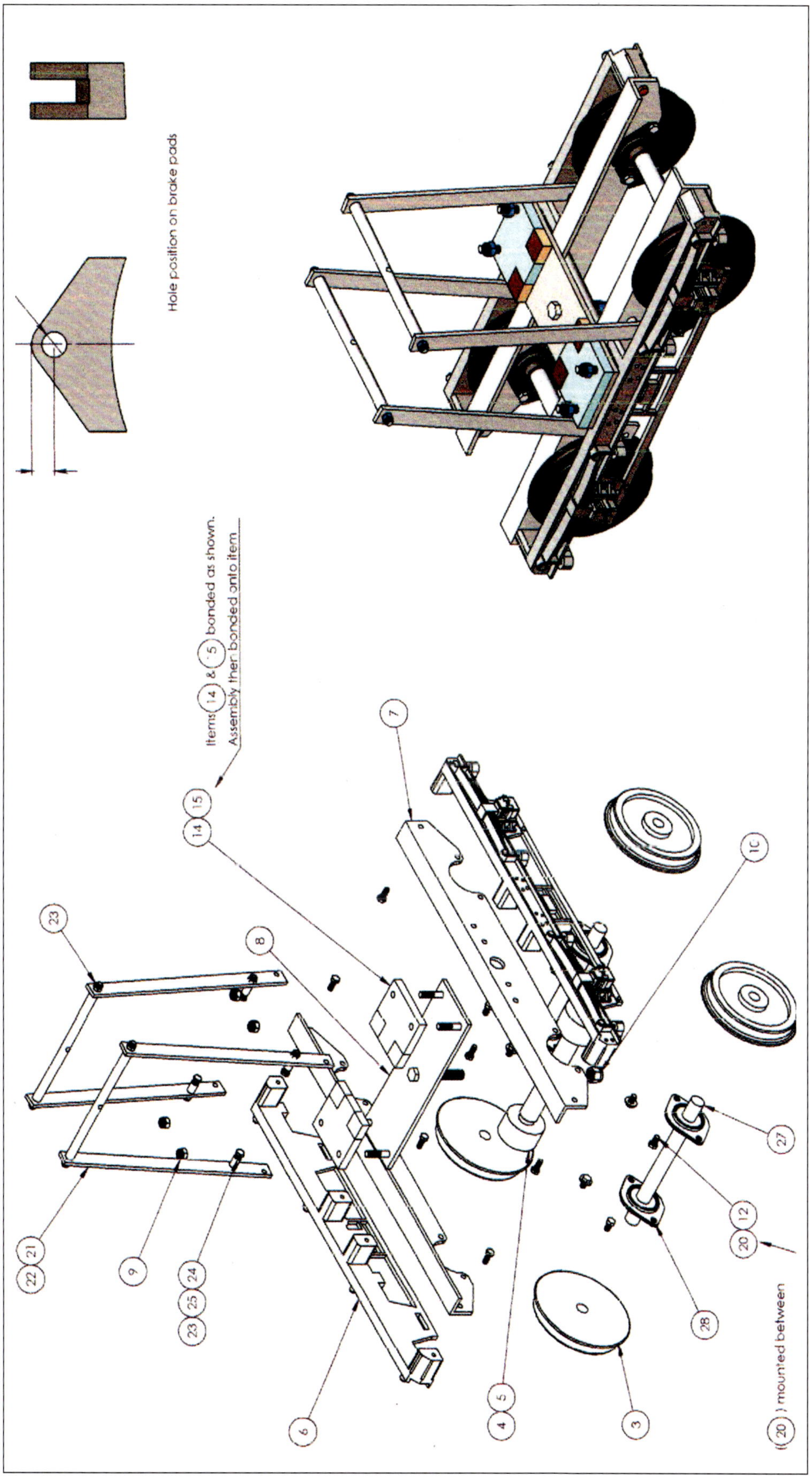

Aristocraft drawing of the exploded bogie.

A fully loaded train formed of Aristocraft coaches on Ted Martin's railway.

to ensure that buffers do not interlock on curves as this will lead to derailments. I have found the safest way to avoid this problem is to connect the coaches with a fixed bar. Trial and error will quickly determine the shortest safe distance between adjacent vehicles. I originally used chain to connect adjacent coaches, but this did lead to derailments due to buffer locking, when propelling.

12.3 Commercial Coaches

For those in a hurry, several commercial companies supply ready to run vehicles of a wide variety of types. Having purchased one second hand Aristocraft coach, I intend to purchase more of these as they run beautifully, look great and are trouble free. The photos below show the quality of some of their range of rolling stock.

Aristocraft Pullman coach.

Rake of LMS Aristocraft coaches.

Dave Owen proudly in charge of Chris Knibbs 4701 and rake of Aristocraft GWR coaches.

A rake of teak Aristocraft coaches.

CHAPTER 13

THE FUTURE OF MODEL ENGINEERING AND THE DEVELOPMENT OF TOMORROW'S ENGINEERS

Model engineering has now existed as a hobby for over 100 years so there is no doubt that this activity has a unique appeal and charm to a lot of people.

During the early 1900s and for a significant part of the last century, Great Britain was a respected and successful manufacturing country from which its wealth was generated. Machine shops were plentiful and therefore a demand existed for skilled machinists which were also plentiful. Industry recognised that they needed to ensure a constant supply of such skilled people and therefore companies offered apprenticeships to train and retain school leavers. Even in the 1960s, most schools had a metalwork shop and here students first learned the basic skills of using hand tools before progressing on to lathes and milling machines. This was my first introduction to metalwork and the *Model Engineer Magazine* – happy days!

As the world has become more computer focussed and automated, workshop machinery has also had to follow this trend to remain competitive. Nowadays, schools and even colleges no longer feature well equipped workshops. Many employers have ceased to offer apprenticeships which has led to the current situation of a national shortage of engineers and skilled machinists.

Over the last fifty years, the heritage world has been fortunate to attract so many people willing to support and carry on those time-honoured traditions and skills, largely on a voluntary basis. These engineers came directly from working in the steam era and provided first hand training to the next generation of enthusiasts. Several generations later, it is becoming increasingly difficult to attract people with the same interests, passions, and skills.

Many model engineering clubs are working hard to encourage young people into the hobby by offering appropriate training. One company has gone a step further in trying to provide training and experiences within the rail industry. 'Train4Rail' (pjd@Train4Rail.co.uk) are committed to educating young would be engineers by providing them with real life experiences within a safe and controlled environment. This is often achieved using miniature railways. Students are given hands-on experiences of all aspects of railway engineering generating safety awareness, communication skills and practical knowledge of building and operating a miniature railway.

They operate on several heritage railway sites and their work neatly integrates with the full-size railway running alongside. With the support of the national rail industry as well as other industries and model engineering clubs, such schemes can greatly help develop young students who might wish to become the engineers of tomorrow.

The rail industry is now one of the largest employers of technical staff in the country and this is in stark contrast to the 1970s when many of its routes and workshops were closing. Modern railways are using the state-of-the-art technology and therefore offer a rewarding career in the industry. The perception that all the work is heavy and dirty is also largely out of date.

The Institute of Mechanical Engineers (IMechE) are also working hard to improve the appeal of a career in engineering and, together with initiatives from other bodies and colleges, students can be trained to meet the needs of the sector. It would be wonderful to see the re-establishment of train building workshops in this country to help revive British Manufacturing – after all, Great Britain is the home of the train!

Let's hope that model engineering can contribute to resolving the desperate shortage of competent professional engineers in our country.

Thomas driving a freight train.

Young driver oiling up at Wansford.

Wren (3 years old) takes the controls of the 08.

Keavy McBrown checks the crossing on the Bentley Miniature Railway.

The Wansford Miniature Railway.

Eva and Anna enjoying a drive.

A young driver at the controls on Jim Ewin's track.

Driver training at the Wansford Miniature Railway.

Family fun at Jim's Railway with the B4 and Paul.

Pat and Michael say that they have finally found a 'decent trainset' after 70 years of looking!

The Bentley Miniature Railway team at the Uckfield Model Railway Show forging the link with railway modellers.

7823 ready for the day ahead.

7823 departing with the evening Cambrian Coast Express.

CONCLUSION

The childhood joy of seeing a steam locomotive has remained with me throughout my entire career in railways. Their enormous energy seems to stimulate all the senses by their smell, sound, and touch. Almost without exception, young and old alike turn their heads and stare as a steam engine pulls away or goes rushing past. It's truly poetry in motion, and I love it.

My overwhelming desire in writing this book has been to share these pleasures and to encourage and inspire would be engineers, model engineers and all those interested in creating anything, to simply decide to create your dream and have a go.

Having the courage to dare to start is usually the most challenging part. There will always be unknowns and uncertainties, but by confronting them, one at a time, my experience is that a way forward emerges and, before you know it, you have succeeded.

Being creative, within the railway world or any other, helps to develop life skills and a self-confidence which will carry through to everything else you do.

My journey into the world of model engineering has brought me a huge amount of fun, knowledge, good friends, a satisfaction of achievement and a practical, pragmatic approach to problem solving.

In an age where fewer people have been trained in a craft and even less have any interest in getting involved with practical work, the hobby of model engineering offers a vital lifeline for the preservation of basic craft skills, knowledge, and a 'can-do' mentality.

Remember that you need not be alone in your battles with making tricky components. Joining a club will prove an endless source of advice, information, enjoyment, and good humour. All club members will have started from a similar place and would be pleased to help newcomers as once they were helped. Their experiences and suggestions can save a lot of time and frustration.

Most importantly enjoy the journey and share the pleasures along the way!

Happy Model Engineering

Paul J. Carpenter C.Eng. BSc (Hons)
FIMechE

ACKNOWLEDGEMENTS

I would like to record my sincere and grateful thanks to the following who have encouraged, supported and contributed to the production of this book:

Simon Bowditch – I cannot overstate the value of Simon's contribution to this book. He has had the patience of a saint and provided us all with great insights into the 'Black Art' of valve gears. His work will significantly advance the quality of all future published designs by giving clarity and understanding to the great importance of attention to detail. I sincerely hope he can be persuaded to follow on now with his own book, full of a wealth of gems and knowledge,

Richard Linkins – Richard and I have been friends for some 50 years and I have witnessed firsthand the beautiful locos he has built that always work perfectly. He has generously shared his insights into the finer points of injectors with us and his writings and lectures have certainly encouraged me to have a go at making my own. In true LBSC style, he makes it all sound so achievable and is ever available with help, suggestions and advice.

Bob Underwood – Bob is a quiet and supportive great friend who is always first on the scene if anyone has a problem. He is encouraging and enthusiastic to everyone and is an endless source of original and great ideas.

Pat Harris is a lifelong friend, who is building a magnificent Black 5 in 7¼in gauge. He has generously allowed his superb workmanship to be presented here.

Colin Driver is an outstanding CAD draughtsman who has painstakingly

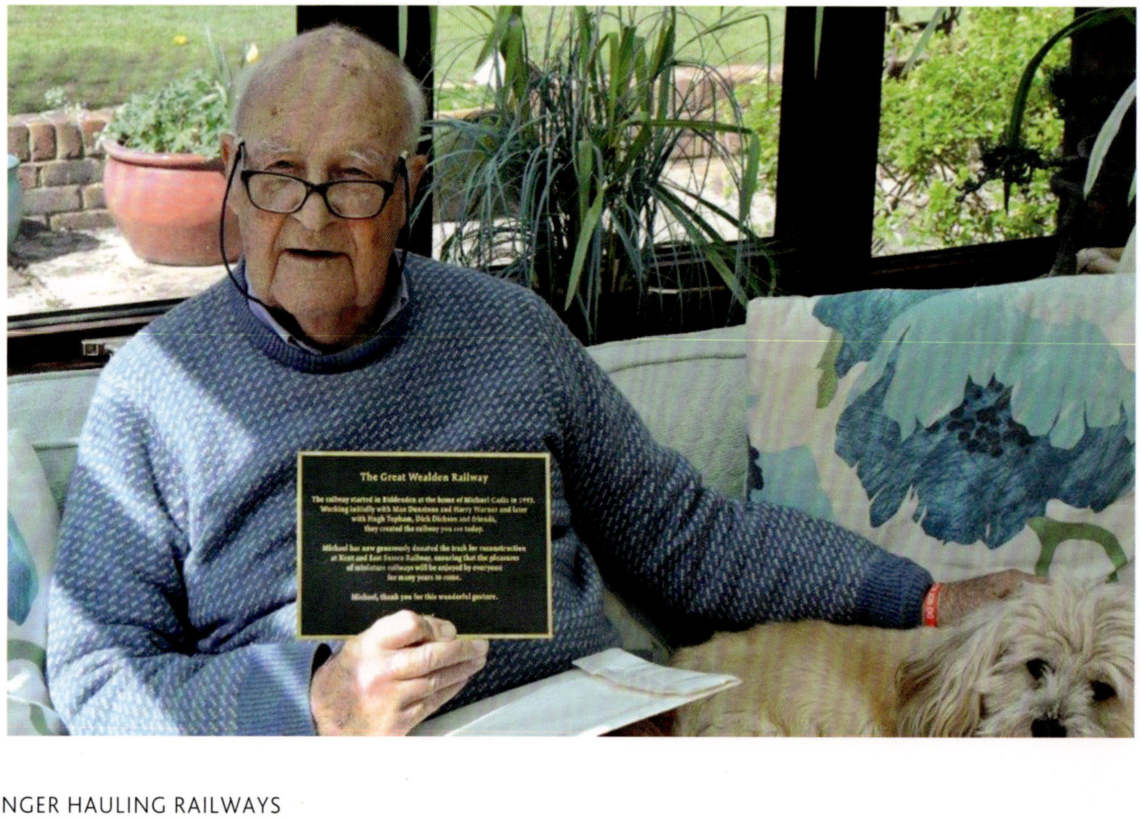

Michael Cadiz with the plaque acknowledging his gift of most of the Great Wealden Railway track to the KESR which is to be re-built on their heritage line.

interpreted my rough sketches into drawings for the book.

Don Ashton – an inspiration to us all. He would be delighted that his work was being used to improve loco designs for the future.

Jim Ewins – a wonderful mentor and friend.

Paul Parkinson – for allowing his expert and beautiful photos to be presented here.

Pam Sage – For allowing me access to Jim's writings and works of the master.

Clare Carpenter – my wife – for her endless patience with me whilst writing this book, her helpful suggestions and her great skill with proof reading.

Michael and Carla Cadiz – real kindred spirits, great friends and generous supporters in helping to complete my railway.

Hugh Topham – for working so hard and enthusiastically to complete the railway and building all those sidings!

John 'Jack' Carpenter – my Dad – the man who made everything possible for me. I was blessed with great and supportive parents and sister Linda. Not much would have happened without them.

Finally, I would like to thank all the many model engineering friends I have met up and down the country over many years for their support and encouragement.

Dad, Mum and sister Linda on a very special day.

APPENDIX 1

DESIGN SIMULATORS BY SIMON BOWDITCH

Both the Bill Hall and Dockstader valve gear simulation programs work on Microsoft Windows. They continue to perform well on my Windows10 laptop.

I had hoped to illustrate this appendix with screenshots of each stage of the simulation process described in the text. Unfortunately, the low-definition graphics generated by both the Bill Hall and the Dockstader simulators belong to the periods when the programs were written – in the case of Bill Hall it was 2001 and those of Charlie Dockstader's date from 2007. I am sure you will have no trouble viewing them on your computer screen, but the fact that the images pixelate on enlargement renders them unsuitable for printing.

Please note that both simulators use the centre line of the motion and the centre of the driving axle as their datum line. This differs from many sets of construction drawings sold by various model engineering suppliers where the centre of the weighshaft may be given as dimensions taken from the edge of the frame etc.

1.1 Using Professor Hall's Program

When you download Bill Hall's simulation programs for Stephenson and Walschaerts valve gear, please ensure you include his written manual entitled: *WALSCHAERTS & STEPHENSON VALVEGEAR PROGRAMMES* (Windows Versions). This six-page document provides the basic information for getting you started with this simulator. Inevitably, a document of only 2,700 words is too brief for the author to support you fully in the exploration of the full potential of the program, but it will help you get started. The data required is very specific e.g. it is essential that you differentiate between the radius and the diameter of a circle. To further assist, Bill Hall provides a simple outline interactive drawing where the item for which data is being entered turns from green to red:

1.2 Using the Dockstader Program

When you download Charlie Dockstader's software, make sure you have included his document entitled *VALVE GEAR PROGRAM INSTRUCTIONS RUNNING IN MICROSOFT WINDOWS*. This five-page document totalling just over 2,800 words gives you all the instructions needed to understand the generic dialogue box which is used by all 67 different valve gear simulations available. Please ensure you are looking at the correct valve type and that you understand the difference between outside admission (slide valve) and inside admission (usually piston valve) when selecting a program. Each of the valve gear simulations offered has its own set of instructions, some of which are more comprehensive than others. All of them, having originated in the USA, use that country's terminology for the various valve gear components.

Appendices 1.1 & 1.2 are not a comprehensive guide to every aspect of valve gear simulation and optimisation, but I hope that it will encourage you to download the free software, install it on your computer and start experimenting. I hope that the simulations that appear in the following section of this book will give you a good starting point from which to explore either Bill Hall's or Charlie Dockstader's software.

APPENDIX 2

DATA INPUT AND VALVE EVENT TABLES

The following data and valve event tables provide evidence of valve gear performance, in both its original and modified geometries for the twelve locomotive designs in the four gauges of Chapter 6.

The file numbers correspond to those of the engine being studied. Three tables are provided for each design i.e., the data input table, the original valve events table and the modified valve events table.

2.5in Gauge

2.1 *Purley Grange*

Dockstader Simulation of Outside Admission GWR Stephenson Valve Gear			
Locomotive: 2.5″ Gauge LBSC "Purley Grange" 4-6-0			
Frame Dimensions		**Purley Grange Original**	**Purley Grange Don Ashton**
Driver	*Diameter*	2.875″	
Stroke	*Length*	1.125″	
Main Rod	*Length*	5.125″	
Eccentric Circle	*Diameter*	0.250″	
Eccentric	*Diameter*	0.875″	
Eccentric Angle	*Forward*	255.52	241.836
Eccentric Angle	*Backward*	255.52	241.836
Eccentric Rod	*Forward*	2.125″	
Eccentric Rod	*Backward*	2.125″	
Link	*Radius*	2.375″	
Link Centre Pivot	*Back Set*	0.000″	0.142″
Link Centre Pivot	*Vertical*	0.3125″	
Link Eccentric Rod Pivot	*Back Set*	0.250″	
Link Block Hanger	*Length*	0.625″	
Hanger Pivot	*Vertical*	0.625″	
Hanger Pivot	*Horizontal*	2.375″	
Valve Extension Rod	*Length*	3.188″	
Rocker Upper Length	*Vertical*	0.625″	

Dockstader Simulation of Outside Admission GWR Stephenson Valve Gear			
Locomotive: 2.5″ Gauge LBSC "Purley Grange" 4-6-0			
Frame Dimensions		**Purley Grange Original**	**Purley Grange Don Ashton**
Rocker Lower End	*Vertical*	0.000″	
Rocker Lower	*Back Set*	0.000″	
Rocker Pivot	*Vertical*	1.500″	
Rocker Pivot	*Horizontal*	5.438″	
Valve Connecting Rod	*Length*	0.438″	
Reversing Arm Vertical	*Length*	0.750″	
Reversing Arm	*Back Set*	0.000″	
Reverse Lifting Arm	*Length*	0.813″	**0.625″**
Reverse Lifting Link	*Length*	1.000″	
Reversing Lift Arm Pivot	*Vertical*	1.000″	
Reversing Lift Arm Pivot	*Horizon*	1.625″	
Reverse Arm to Centre	*Normal*	0.375″	**0.391″**
Reverse Arm to Centre	*Maximum*	0.438″	**0.407″**
Cylinder Dimensions			
Cylinder centre to Driving centre	*Length*	7.813″	
Cylinder Bore	*Diameter*	0.813″	
Cylinder Bore	*Length*	1.625″	
Cylinder Steam Port	*Width*	0.063″	
Piston Centre to Cross Head Centre		2.688″	
Piston Width		0.375″	
Valve Surface to Cylinder CL		0.625″	
Valve Stem CL to Cylinder CL		0.875″	
Exhaust Port	*Width*	0.250″	
Bridge Width	*Rear*	0.094″	
Bridge Width	*Front*	0.094″	
Steam Port Width	*Rear*	0.094″	
Steam Port Width	*Front*	0.094″	
Valve Chamber	*Height*	0.437″	
Valve Chamber	*Length*	1.250″	
Valve Height		0.375″	
Valve Length	*Rear Half*	0.375″	
Valve Length	*Front Half*	0.375″	
Valve Exhaust Cavity	*Rear Half*	0.227″	**0.219″**
Valve Exhaust Cavity	*Front Half*	0.277″	**0.219″**
Valve Centre to Joint	*(Valve Stem)*	1.993″	**1.960″**

Valve events from Stephenson's Valve Gear fitted to 2.5″ Gauge 4-6-0 Tender Loco (LBSC's Original Design of "Purley Grange")														
Notch	Angle	Travel	Compression		Admission		Cut-Off		Exhaust		Port Opening		Lead	
			Front	Back	Front	Back	Front	Back	Front	Back	Front	Back	Front	Back
0.8	1.879	0.222″	0.0%	6.8%	12.8%	0.1%	73.0%	86.2%	89.4%	98.8%	0.025″	0.073″	-0.055″	0.007″
0.7	1.848	0.199″	0.1%	8.6%	15.4%	0.4%	64.1%	81.5%	87.3%	97.9%	0.013″	0.062″	-0.053″	0.010″
0.6	1.817	0.177″	0.1%	10.6%	27.1%	1.0%	47.3%	75.5%	82.5%	97.9%	0.001″	0.052″	-0.051″	0.012″
0.5	1.785	0.156″	0.5%	15.0%		1.0%		68.3%	77.1%	95.2%	-0.010″	0.042″	-0.049″	0.014″
0.4	1.754	0.137″	1.2%	20.2%		1.7%		58.2%	71.1%	93.5%	-0.021″	0.034″	-0.047″	0.016″
0.3	1.723	0.121″	2.1%	25.9%		2.7%		44.7%	61.4%	89.5%	-0.031″	0.028″	-0.046″	0.017″
0.2	1.692	0.109″	3.3%	35.3%		35.3%		31.6%	47.5%	84.6%	-0.038″	0.023″	-0.045″	0.018″
0.1	1.660	0.101″	8.4%	49.0%		49.0%		20.1%	30.3%	72.9%	-0.043″	0.020″	-0.045″	0.019″
0	1.629	0.100″	63.0%	18.1%		12.7%		12.6%	12.8%	59.5%	-0.043″	0.019″	-0.045″	0.019″
-0.1	1.598	0.101″	49.0%	6.5%		8.6%		19.6%	30.3%	72.9%	-0.043″	0.020″	-0.045″	0.019″
-0.2	1.567	0.108″	38.6%	3.3%		5.2%		31.1%	47.5%	84.6%	-0.038″	0.022″	-0.045″	0.018″
-0.3	1.535	0.120″	28.9%	2.1%		3.9%		44.7%	57.9%	89.5%	-0.031″	0.027″	-0.046″	0.018″
-0.4	1.504	0.134″	1.2%	20.2%		1.7%		58.3%	67.9%	93.5%	-0.023″	0.033″	-0.047″	0.016″
-0.5	1.473	0.149″	0.5%	15.0%		1.7%		68.5%	77.1%	96.7%	-0.015″	0.040″	-0.049″	0.015″
-0.6	1.442	0.168″	0.1%	12.7%		1.0%		75.7%	82.5%	97.9%	-0.005″	0.049″	-0.051″	0.013″
-0.7	1.410	0.186″	0.1%	10.6%	20.9%	0.4%	54.0%	81.7%	85.0%	97.9%	0.004″	0.058″	-0.053″	0.011″
-0.8	1.379	0.206″	0.0%	6.8%	15.4%	0.4%	66.8%	86.2%	89.4%	98.8%	0.014″	0.068″	-0.055″	0.008″
Port Width Front Port Width Rear			0.094″ 0.094″	Lap of Valve Front Lap of Valve Rear			0.062″ 0.062″	Exhaust Clearance Front Rear			0.008″ 0.008″	Outside Admission GWR Stephenson's Simulation Dockstader 04/01/2023		
Link Centre Pivot Back Set			**0.000″**	Angle of Advance of Eccentric			**14.48°**	Length of Lifting Arm			**0.813″**			

Valve events from Stephenson's Valve Gear fitted to 2.5" Gauge 4-6-0 Tender Loco (Don Ashton Version of "Purley Grange")														
Notch	Angle	Travel	Compression		Admission		Cut-Off		Exhaust		Port Opening		Lead	
			Front	Back	Front	Back	Front	Back	Front	Back	Front	Back	Front	Back
0.8	2.017"	0.263"	6.5%	5.2%	0.0%	0.0%	79.7%	77.6%	94.8%	93.5%	0.065"	0.074"	-0.003"	-0.003"
0.7	1.968"	0.240"	10.5%	8.6%	0.1%	0.1%	75.4%	72.0%	91.4%	89.5%	0.054"	0.062"	0.001"	0.001"
0.6	1.919"	0.219"	12.8%	10.6%	0.1%	0.1%	67.3%	65.4%	89.4%	87.2%	0.044"	0.051"	0.004"	0.004"
0.5	1.870"	0.200"	15.4%	15.0%	0.5%	0.4%	59.7%	58.1%	85.0%	84.6%	0.035"	0.041"	0.007"	0.007"
0.4	1.822"	0.184"	20.9%	17.5%	1.2%	1.0%	49.4%	48.5%	82.5%	79.1%	0.028"	0.032"	0.010"	0.009"
0.3	1.773"	0.170"	27.1%	25.9%	2.1%	1.7%	36.5%	38.0%	74.1%	72.9%	0.022"	0.024"	0.012"	0.010"
0.2	1.724"	0.159"	33.6%	32.1%	3.3%	2.7%	26.0%	27.8%	67.9%	66.4%	0.017"	0.018"	0.013"	0.012"
0.1	1.675"	0.153"	44.0%	42.1%	6.5%	5.2%	17.0%	15.9%	57.9%	56.0%	0.015"	0.014"	0.014"	0.013"
0	1.626"	0.151"	51.0%	52.5%	10.5%	8.6%	9.9%	9.9%	47.5%	49.0%	0.014"	0.013"	0.014"	0.013"
-0.1	1.577"	0.153"	40.5%	42.1%	6.5%	5.2%	15.2%	16.7%	57.9%	59.5%	0.015"	0.014"	0.014"	0.013"
-0.2	1.528"	0.158"	33.6%	32.1%	3.3%	2.7%	23.7%	25.6%	67.9%	66.4%	0.017"	0.017"	0.013"	0.012"
-0.3	1.479"	0.168"	27.1%	25.9%	2.1%	1.7%	35.6%	37.6%	74.1%	72.9%	0.021"	0.023"	0.012"	0.011"
-0.4	1.431"	0.180"	20.9%	20.2%	1.2%	1.0%	46.1%	47.9%	78.8%	79.1%	0.026"	0.030"	0.010"	0.009"
-0.5	1.386"	0.194"	15.4%	15.0%	0.5%	0.4%	56.3%	57.7%	85.0%	84.6%	0.032"	0.038"	0.008"	0.007"
-0.6	1.333"	0.212"	12.8%	10.6%	0.1%	0.1%	66.3%	65.1%	89.4%	87.2%	0.039"	0.049"	0.005"	0.004"
-0.7	1.284"	0.230"	10.5%	8.6%	0.1%	0.1%	72.9%	71.7%	91.4%	89.5%	0.047"	0.059"	0.001"	0.000"
-0.8	1.235"	0.250"	6.5%	5.2%	0.0%	0.0%	78.8%	77.4%	94.8%	93.5%	0.056"	0.070"	-0.003"	-0.004"

Port Width Front Port Width Rear	0.094" 0.094"	Lap of Valve Front Lap of Valve Rear	0.062" 0.062"	Exhaust Clearance Front Rear	0.000" 0.000"	Outside Admission GWR Stephenson's Simulation Dockstader 06/01/2023	
Link Centre Pivot Back Set	**0.142"**	Angle of Advance of Eccentric	**28.16°**	Length of Lifting Arm	**0.625"**		

Compression Score	66 / 68	Admission Score	68 / 68	Cut-Off Score	66 / 68	Exhaust Score	66 / 68	Total Score	266 / 272
Compression %	97.1%	Admission %	100%	Cut-Off %	97.1%	Exhaust %	97.1%	Total %	97.8%

2.2 *Princess Royal*

Input Data for Dockstader's Outside Admission Walschaert's Valve Gear 2.5" Gauge LMS "Princess Royal" Class 4-6-2 Locomotive by LBSC.				
Item	**Description**		**Lizzie One**	**Lizzie Two**
Frame Dimensions				
1	Driving Wheel	Diameter	3.5625″	3.563″
2	Stroke	Length	1.125″	1.125
3	Connecting Rod	Length	4.813″	4.813″
4	Drop Link to Crosshead	Vertical	0.563″	**0.594″**
5	Drop Link to Crosshead	Backset	0.000″	0.000″
6	Union Link	Length	0.625″	**0.632″**
7	Combination Link	Upper	0.188″	0.188″
8	Combination Link	Backset	0.000″	0.000″
9	Combination Link	Lower	1.312″	1.312″
10	Radius Rod	Length	2.313″	2.313″
11	Expansion Link	Radius	2.313″	2.313″
12	Expansion Link Crank	Vertical	0.498″	**0.537″**
13	Expansion Link Crank	Backset	-0.063	**0.005″**
14	Eccentric Rod	Length	3.034″	**2.972″**
15	Eccentric Crank Circle	Diameter	0.375″	0.375″
16	Eccentric Crank	Length	0.5769″	**0.578″**
17	Radius Rod Extension	Length	0.500″	0.500″
18	Reverse Arm	Vertical	1.000″	1.000″
19	Reverse Arm	Backset	0.047″	0.000″
20	Lifting Arm	Length	0.656″	0.656″
21	Lifting Link	Length	0.000″	0.000″
22	Reverse Arm Pivot	Vertical	0.708″	**0.786″**
23	Reverse Arm Pivot	Horizontal	1.975″	1.975″
24	Expansion Link Pivot	Vertical	0.785″	**0.749″**
25	Expansion Link Pivot	Horizontal	2.973″	2.973″
26	Reverse Arm to Centre Normal		0.531″	0.531″
27	Reverse Arm to Centre Maximum		0.594″	0.594″

Input Data for Dockstader's Outside Admission Walschaert's Valve Gear 2.5″ Gauge LMS "Princess Royal" Class 4-6-2 Locomotive by LBSC.			
Item	**Description**	**Lizzie One**	**Lizzie Two**
Cylinder Dimensions			
28	Cylinder Centre to Driver Centre	7.453″	7.453″
29	Cylinder Centre above Driver	n/a	n/a
30	Cylinder Bore Diameter	0.6875″	0.6875″
31	Cylinder Bore Length	1.684″	1.684″
32	Cylinder Steam Port Width	0.063″	0.063″
33	Piston Centre to Crosshead Centre	2.625″	2.625″
34	Piston Width	0.438″	0.438″
35	Valve Surface to Cylinder C/L (Vertical)	0.625″	0.625″
36	Valve Stem C/L to Cylinder C/L (Vertical)	0.875″	0.875″
37	Exhaust Port Width	0.188″	0.188″
38	Bridge Width (Rear)	0.094″	0.094″
39	Bridge Width (Front)	0.094″	0.094″
40	Steam Port Width (Rear)	0.094″	0.094″
41	Steam Port Width (Front)	0.094″	0.094″
42	Valve Chamber Height	0.500″	0.500″
43	Valve Chamber Length	1.250″	1.250″
44	Valve Height	0.438″	0.438″
45	Valve Length (Rear Half)	0.3438″	0.345″
46	Valve Length (Front Half)	0.3438″	0.345″
47	Valve Exhaust Cavity (Rear Half)	0.1953″	0.187″
48	Valve Exhaust Cavity (Front Half)	0.1953″	0.190″
49	Valve Centre to Joint (Valve Stem)	2.098″	2.189″

Valve events from Walschaert's Valve Gear fitted to 2.5″ Gauge LM&SR "Princess" Class 4-6-2 by LBSC														
Notch	Angle	Travel	Compression		Admission		Cut-Off		Exhaust		Port Opening		Lead	
			Front	Back	Front	Back	Front	Back	Front	Back	Front	Back	Front	Back
0.8	28.0°	0.294″	8.063%	6.975%	0.641%	0.503%	76.13%	75.44%	90.38%	89.14%	0.079″	0.091″	0.017″	0.017″
0.7	24.5°	0.270″	9.588%	8.375%	0.787%	0.618%	71.86%	71.49%	88.51%	87.18%	0.068″	0.078″	0.017″	0.017″
0.6	21.0°	0.246″	11.60%	10.25%	1.006%	0.789%	66.39%	66.48%	86.04%	84.63%	0.057″	0.065″	0.017″	0.017″
0.5	17.5°	0.222″	14.28%	12.80%	1.331%	1.046%	59.42%	60.10%	82.75%	81.29%	0.046″	0.052″	0.017″	0.017″
0.4	14.0°	0.201″	17.87%	16.27%	1.852%	1.461%	50.77%	52.00%	78.36%	76.92%	0.036″	0.041″	0.017″	0.017″
0.3	10.5°	0.183″	22.65%	20.98%	2.720%	2.159%	40.59%	41.92%	72.60%	71.23%	0.028″	0.031″	0.017″	0.017″
0.2	7.0°	0.169″	28.93%	27.18%	4.269%	3.424%	29.71%	30.24%	65.32%	63.96%	0.022″	0.023″	0.017″	0.017″
0.1	3.5°	0.160″	36.89%	34.94%	7.073%	5.765%	19.67%	18.91%	56.67%	55.08%	0.018″	0.018″	0.016″	0.016″
0	0.0°	0.156″	47.51%	45.11%	11.95%	10.59%	11.99%	9.979%	46.32%	43.80%	0.016″	0.016″	0.016″	0.016″
-0.1	-3.5°	0.159″	38.69%	35.30%	7.012%	5.789%	19.69%	16.72%	56.33%	52.87%	0.018″	0.017″	0.016″	0.016″
-0.2	-7.0°	0.166″	31.02%	26.95%	4.208%	3.323%	29.82%	25.67%	65.55%	61.20%	0.021″	0.021″	0.016″	0.016″
-0.3	-10.5°	0.178″	24.82%	20.58%	2.670%	2.058%	40.89%	35.44%	73.07%	68.26%	0.027″	0.027″	0.016″	0.016″
-0.4	-14.0°	0.193″	20.00%	15.96%	1.809%	1.371%	51.18%	44.63%	78.77%	73.93%	0.035″	0.034″	0.016″	0.016″
-0.5	-17.5°	0.210″	16.33%	12.66%	1.295%	0.971%	59.69%	52.52%	82.95%	78.38%	0.044″	0.042″	0.016″	0.016″
-0.6	-21.0°	0.227″	13.55%	10.29%	0.977%	0.728%	66.35%	59.00%	86.01%	81.84%	0.053″	0.050″	0.015″	0.015″
-0.7	-24.5°	0.246″	11.45%	8.563%	0.766%	0.571%	71.44%	64.23%	88.28%	84.50%	0.063″	0.059″	0.015″	0.015″
-0.8	-28.0°	0.263″	9.840%	7.286%	0.622%	0.459%	75.32%	66.42%	89.97%	86.57%	0.072″	0.067″	0.015″	0.015″
Width of Steam Port			0.094″	Lap of Valve			0.062″	Exhaust Clearances			0.007″	Prof. Bill Hall's Walschaerts' Valve Gear Simulation		

Compression Score	57/68	Admission Score	68/68	Cut-Off Score	46/68	Exhaust Score	53/68	Total 224/272
Percentage Score	83.8%	Percentage Score	100%	Percentage Score	67.6%	Percentage Score	77.9%	Overall 82.3%

Valve events from Walschaert's Valve Gear fitted to 2.5″ Gauge LM&SR "Princess" Class 4-6-2 (MK2) by LBSC														
Notch	Angle	Travel	Compression		Admission		Cut-Off		Exhaust		Port Opening		Lead	
			Front	Back	Front	Back	Front	Back	Front	Back	Front	Back	Front	Back
0.8	2.562″	0.296″	8.4%	8.5%	0.5%	0.4%	77.2%	77.0%	91.5%	89.4%	0.081″	0.091″	0.014″	0.016″
0.7	2.495″	0.271″	10.6%	10.5%	0.5%	1.0%	73.0%	72.2%	89.5%	89.4%	0.069″	0.078″	0.014″	0.016″
0.6	2.429″	0.247″	12.9%	12.6%	1.2%	1.0%	67.2%	67.7%	87.4%	87.1%	0.058″	0.065″	0.014″	0.016″
0.5	2.362″	0.224″	15.4%	14.9%	1.2%	1.0%	60.8%	61.2%	85.1%	84.6%	0.047″	0.053″	0.014″	0.016″
0.4	2.296″	0.202″	21.0%	20.0%	1.2%	1.7%	53.1%	54.0%	80.0%	79.0%	0.036″	0.042″	0.014″	0.016″
0.3	2.229″	0.182″	24.1%	25.7%	2.2%	2.7%	42.7%	43.8%	74.3%	75.9%	0.027″	0.031″	0.014″	0.016″
0.2	2.163″	0.168″	30.4%	31.9%	3.4%	3.8%	30.3%	31.4%	68.1%	66.2%	0.021″	0.023″	0.014″	0.016″
0.1	2.096″	0.158″	40.7%	41.9%	6.5%	5.2%	20.2%	19.8%	58.1%	59.3%	0.016″	0.018″	0.014″	0.016″
0	2.030″	0.154″	54.7%	52.3%	12.9%	12.6%	10.2%	9.9%	47.7%	45.3%	0.014″	0.016″	0.014″	0.016″
-0.1	1.963″	0.158″	44.2%	41.9%	6.5%	6.8%	17.8%	16.7%	58.1%	55.8%	0.016″	0.018″	0.014″	0.016″
-0.2	1.897″	0.166″	33.8%	31.9%	3.4%	3.8%	29.3%	27.4%	64.9%	66.2%	0.020″	0.022″	0.014″	0.016″
-0.3	1.830″	0.179″	27.2%	25.7%	2.2%	2.7%	40.5%	37.4%	74.3%	72.8%	0.027″	0.028″	0.014″	0.017″
-0.4	1.764″	0.197″	21.0%	20.0%	2.2%	1.2%	53.0%	47.5%	80.0%	79.0%	0.036″	0.037″	0.014″	0.017″
-0.5	1.697″	0.217″	18.2%	14.9%	1.2%	1.0%	61.2%	55.7%	85.1%	81.8%	0.046″	0.047″	0.014″	0.017″
-0.6	1.631″	0.239″	15.4%	12.5%	1.2%	1.0%	69.4%	64.2%	87.4%	84.6%	0.058″	0.057″	0.014″	0.017″
-0.7	1.564″	0.263″	12.9%	10.5%	0.5%	1.0%	73.9%	68.5%	89.5%	87.1%	0.070″	0.069″	0.014″	0.017″
-0.8	1.498″	0.287″	10.6%	8.5%	0.5%	0.4%	78.4%	73.9%	91.5%	89.5%	0.083″	0.080″	0.014″	0.017″
Width of Steam Port			0.094″	Lap of Valve			0.062″	Exhaust Clearances			0.000″	Charlie Dockstader's Walschaerts' Valve Gear Simulation		
Compression Score		62/68	Admission Score		68/68	Cut-Off Score		53/68	Exhaust Score		61/68	Total 244/272		
Percentage Score		91.1%	Percentage Score		100%	Percentage Score		77.9%	Percentage Score		89.7%	Overall 89.7%		

3.5in Gauge

2.3 *Lilla*

Dockstader Simulation of Outside Admission Stephenson Valve Gear				
Locomotive: 3.5" R. Harrison. 0-4-0 Saddle Tank "Lilla" 22nd March 2023				
Frame Dimensions		**Lilla Original**	**Lilla "C" (Loco Link)**	**Lilla "Mk 3" (Launch Link)**
Driver	*Diameter*	3.750"		
Stroke	*Length*	2.000"		
Main Rod	*Length*	8.500"		
Eccentric Circle	*Diameter*	0.500"	0.625"	0.500"
Eccentric	*Diameter*	1.375"		
Eccentric Angle	*Forward*	240.700°	254.300°	248.000°
Eccentric Angle	*Backward*	240.700°	254.300°	248.000°
Eccentric Rod	*Forward*	4.250"	4.750"	4.250"
Eccentric Rod	*Backward*	4.250"	4.750"	4.250"
Link	*Radius*	4.750"	4.750"	4.750"
Link Centre Pivot	*Back Set*	0.000"	-0.063"	0.219"
Link Centre Pivot	*Vertical*	0.000"	0.000"	0.000"
Link Eccentric Rod Pivot	*Vertical*	0.688"	0.688"	0.563"
Link Eccentric Rod Pivot	*Back Set*	0.500"	0.000"	0.500"
Rocker Upper Vertical	*Length*	-50.000		
Rocker Upper	*Backset*	0.000"		
Rocker Lower	*Length*	50.000"		
Rocker Pivot	*Vertical*	50.000"		
Rocker Pivot	*Horizontal*	4.750"	4.750"	5.391"
Valve Connecting Rod	*Length*	6.125"		
Lifting Link	*Length*	1.375"	1.507"	1.633"
Reversing Arm Vertical	*Length*	1.375"	1.216"	1.375
Reversing Arm	*Backset*	0.375"	0.375"	0.219"
Lifting Arm	*Length*	1.375"	1.281"	0.868"
Reversing Lift Arm Pivot	*Vertical*	1.750"	1.750"	1.841"
Reversing Lift Arm Pivot	*Horizon*	3.500"	3.500"	3.622"
Reverse Arm to Centre	*Normal*	0.750"	0.438"	1.000"
Reverse Arm to Centre	*Maximum*	0.813"	0.625"	1.032"

Dockstader Simulation of Outside Admission Stephenson Valve Gear				
Locomotive: 3.5" R. Harrison. 0-4-0 Saddle Tank "Lilla" 22nd March 2023				
Frame Dimensions		**Lilla Original**	**Lilla "C" (Loco Link)**	**Lilla "Mk 3" (Launch Link)**
Cylinder Dimensions				
Cylinder centre to Driving centre	*Length*	13.500"	13.500"	13.588"
Cylinder Bore	*Diameter*	1.250""	1.250"	1.250"
Cylinder Bore	*Length*	2.875"	2.875"	2.875"
Cylinder Steam Port	*Width*	0.125"	0.125"	0.125"
Piston Centre to Cross Head Centre		5.000"	5.000"	5.078"
Piston Width		0.688"	0.688"	0.688"
Valve Surface to Cylinder CL		0.000"	0.000"	0.000"
Valve Stem CL to Cylinder CL		0.000"	0.000"	0.000"
Exhaust Port	*Width*	0.250"	0.250"	0.250"
Bridge Width	*Rear*	0.250"	0.250"	0.250"
Bridge Width	*Front*	0.250"	0.250"	0.250"
Steam Port Width	*Rear*	0.125"	0.125"	0.125"
Steam Port Width	*Front*	0.125"	0.125"	0.125"
Valve Chamber	*Height*	0.750"	0.750"	0.750"
Valve Chamber	*Length*	2.125"	2.125"	2.125"
Valve Height		0.625"	0.625"	0.625"
Valve Length	*Rear Half*	0.625"	0.610"	0.605"
Valve Length	*Front Half*	0.625"	0.610"	0.605"
Valve Exhaust Cavity	*Rear Half*	0.375"	0.373"	0.372"
Valve Exhaust Cavity	*Front Half*	0.375"	0.378"	0.378"
Valve Centre to Joint	(*Valve Stem*)	2.711"	2.685"	2.759"

Valve events from Stephenson's Valve Gear fitted to 3.5" Gauge 0-6-0 ST "Lilla"

(Dockstader Simulation – Original Design with Launch Link)

Notch	Angle	Travel	Compression		Admission		Cut-Off		Exhaust		Port Opening		Lead	
			Front	Back	Front	Back	Front	Back	Front	Back	Front	Back	Front	Back
0.8	4.258"	0.449"	8.4%	12.6%	-0.1%	1.0%	69.0%	68.4%	87.4%	91.6%	0.080"	0.119"	-0.016"	0.036"
0.7	4.164"	0.425"	10.6%	12.6%	-0.1%	1.7%	63.0%	63.8%	87.4%	89.4%	0.067"	0.108"	- 0.010"	0.040"
0.6	4.070"	0.401"	15.5%	17.4%	-0.0%	1.7%	56.4%	57.1%	82.6%	84.5%	0.054"	0.097"	-0.005"	0.044"
0.5	3.976"	0.377"	18.2%	20.0%	-0.0%	2.7%	49.0%	50.1%	80.0%	81.8%	0.041"	0.086"	-0.001"	0.047"
0.4	3.883"	0.357"	24.1%	22.8%	0.1%	3.8%	38.7%	41.8%	77.2%	75.9%	0.030"	0.077"	0.003"	0.051"
0.3	3.789"	0.339"	30.5%	28.7%	0.5%	5.2%	26.2%	33.6%	71.3%	69.5%	0.020"	0.069"	0.006"	0.054"
0.2	3.695"	0.325"	37.2%	35.1%	1.2%	6.8%	15.0%	26.8%	64.9%	62.8%	0.013"	0.062"	0.008"	0.056"
0.1	3.601"	0.318"	44.2%	45.3%	2.2%	10.5%	7.7%	18.9%	54.7%	55.8%	0.010"	0.058"	0.009"	0.057"
0	3.508"	0.316"	51.2%	48.8%	3.4%	14.9%	5.9%	14.1%	51.2%	48.8%	0.009"	0.057"	0.009"	0.057"
-0.1	3.414"	0.322"	40.7%	41.9%	1.2%	10.5%	12.4%	21.0%	58.1%	59.3%	0.013"	0.059"	0.008"	0.057"
-0.2	3.320"	0.333"	33.8%	35.1%	0.5%	6.8%	22.8%	28.0%	64.9%	66.2%	0.019"	0.064"	0.006"	0.055"
-0.3	3.226"	0.351"	24.1%	25.7%	0.1%	5.2%	35.2%	37.2%	74.3%	75.9%	0.029"	0.072"	0.003"	0.053"
-0.4	3.133"	0.373"	18.2%	22.8%	-0.0%	2.7%	45.8%	47.1%	77.2%	81.8%	0.041"	0.082"	-0.002"	0.049"
-0.5	3.039"	0.402"	12.9%	17.4%	-0.0%	1.7%	56.1%	57.1%	82.6%	87.1%	0.056"	0.096"	-0.007"	0.044"
-0.6	2.945"	0.436"	10.6%	12.6%	-0.1%	1.0%	64.6%	65.0%	87.4%	89.4%	0.073"	0.113"	-0.014"	0.038"
-0.7	2.851"	0.472"	8.4%	10.5%	-0.1%	0.4%	72.2%	72.2%	89.5%	91.6%	0.091"	0.131"	-0.022"	0.030"
-0.8	2.758"	0.513"	4.8%	6.8%	-0.5%	0.4%	78.0%	78.1%	93.2%	95.2%	0.111"	0.152"	-0.031"	0.021"

Port Width Front Port Width Rear	0.125" 0.125"	Lap of Valve Front Lap of Valve Rear	0.125" 0.125"	Exhaust Clearance Front Rear	0.000" 0.000"	Outside Admission Stephenson's Simulation	
Link Centre Pivot Back Set	**0.000"**	Angle of Advance of Eccentric	**29.30°**	Length of Lifting Arm	**1.375"**	Dockstader 28/03/2019	

Compression Score	59/68	Admission Score	44/68	Cut-Off Score	49/68	Exhaust Score	59/68	Total Score	211/272
Compression %	86.8%	Admission %	64.7%	Cut-Off %	72.1%	Exhaust %	86.8%	Total %	77.6%

Valve events from Stephenson's Valve Gear fitted to 3.5" Gauge 0-6-0 ST "Lilla"

(Dockstader Simulation – Mark 3 Modified Version with Launch Link)

Notch	Angle	Travel	Compression		Admission		Cut-Off		Exhaust		Port Opening		Lead	
			Front	Back	Front	Back	Front	Back	Front	Back	Front	Back	Front	Back
0.8	4.727	0.460"	6.5%	6.8%	0.0%	0.1%	80.9%	79.8%	93.2%	93.5%	0.121"	0.129"	-0.007"	0.003"
0.7	4.607	0.435"	6.5%	6.8%	0.0%	0.1%	78.2%	76.8%	93.2%	93.5%	0.108"	0.117"	-0.003"	0.005"
0.6	4.477	0.399"	8.4%	8.5%	0.1%	0.1%	73.6%	71.3%	91.5%	91.6%	0.090"	0.099"	0.002"	0.009"
0.5	4.352	0.363"	12.9%	12.6%	0.1%	0.4%	66.9%	64.5%	87.4%	87.1%	0.072"	0.081"	0.006"	0.012"
0.4	4.227	0.327"	15.5%	14.9%	0.5%	0.4%	57.4%	55.3%	85.1%	84.5%	0.055"	0.062"	0.010"	0.015"
0.3	4.102	0.294"	21.0%	20.0%	1.2%	1.0%	45.7%	43.7%	80.0%	79.0%	0.039"	0.045"	0.013"	0.018"
0.2	3.977	0.269"	27.2%	28.7%	2.2%	1.7%	32.4%	31.5%	71.3%	72.8%	0.027"	0.032"	0.016"	0.020"
0.1	3.852	0.253"	40.7%	38.5%	3.4%	3.8%	16.9%	15.4%	61.5%	59.3%	0.020"	0.023"	0.017"	0.021"
0	3.727	0.250"	51.2%	52.3%	8.4%	10.5%	7.7%	7.8%	47.7%	48.8%	0.018"	0.022"	0.018"	0.022"
-0.1	3.602	0.257"	37.2%	38.5%	3.4%	3.8%	17.3%	18.9%	61.5%	62.8%	0.021"	0.026"	0.017"	0.021"
-0.2	3.477	0.276"	27.2%	28.7%	1.2%	1.7%	32.2%	33.4%	71.3%	72.8%	0.029"	0.037"	0.015"	0.019"
-0.3	3.352	0.305"	21.0%	20.0%	0.5%	1.0%	47.4%	47.3%	80.0%	79.0%	0.042"	0.053"	0.012"	0.017"
-0.4	3.227	0.345"	15.5%	12.6%	0.1%	0.4%	62.8%	60.5%	87.4%	84.5%	0.060"	0.075"	0.009"	0.013"
-0.5	3.102	0.392"	10.6%	8.5%	0.1%	0.1%	72.5%	70.4%	91.5%	89.4%	0.082"	0.100"	0.003"	0.007"
-0.6	2.977	0.444"	6.5%	6.8%	0.0%	0.1%	80.7%	76.9%	93.5%	93.5%	0.106"	0.128"	-0.003"	0.001"
-0.7	2.852	0.501"	4.8%	3.8%	0.0%	0.0%	85.7%	82.6%	96.2%	95.2%	0.133"	0.158"	-0.011"	-0.007"
-0.8	2.727	0.563"	3.4%	2.7%	0.1%	0.0%	90.0%	87.0%	97.3%	96.6%	0.162"	0.191"	-0.020"	-0.017"

Port Width Front Port Width Rear	0.125" 0.125"	Lap of Valve Front Lap of Valve Rear	0.105" 0.105"	Exhaust Clearance Front Rear	-0.003" 0.003"	Outside Admission Stephenson's Simulation Dockstader 28/03/2019
Link Centre Pivot Back Set	**0.219"**	Angle of Advance of Eccentric	**22.0°**	Length of Lifting Arm	**0.868"**	

Compression Score	65/68	Admission Score	67/68	Cut-Off Score	59/68	Exhaust Score	65/68	Total Score	256/272
Compression %	95.6%	Admission %	98.5%	Cut-Off %	86.8%	Exhaust %	95.6%	Total %	94.1%

2.4 *Britannia*

Dockstader Valve Gear Programme			
Data for Inside Admission Walschaert's Valve Gear Simulation (3.5″ Gauge "Britannia")			
Data for Loco Frame Dimensions and Linkage		Britannia	Optimised
1	Driving Wheel Diameter	4.625″	
2	Stroke of Crank	1.750″	
3	Length of Connecting Rod (Main Rod)	8.000″	
4	Drop Link to Crosshead C/L (Vertical)	0.500″	
5	Drop Link to Crosshead C/L (Back-set)	0.000″	
6	Union Link	0.844″	
7	Combination Lever (Upper Portion)	0.313″	
8	Combination Lever (Upper Portion Back-set)	0.000″	
9	Combination Lever (Lower Portion)	1.875″	
10	Radius Rod	4.875″	
11	Link Radius	4.875″	
12	Link Crank (Vertical)	1.082″	
13	Link Crank (Backset)	0.308″	
14	Eccentric Rod	3.769″	
15	Eccentric Crank Circ	0.811″	
16	Eccentric Crank Length	0.977″	
17	Radius Rod Extension Length	-0.688″	
18	Reversing Arm (Vertical)	0.750″	
19	Reversing Arm (Back-set)	-0.078″	
20	Lifting Arm (Length)	1.500″	**1.468″**
21	Lifting Link (Length)	0.875″	
22	Reverse Arm Pivot (Vertical)	2.102″	
23	Reverse Arm Pivot (Horizontal)	3.144″	
24	Link Centre Pivot (Vertical)	1.227″	
25	Link Centre Pivot (Horizontal)	3.955″	
26	Reverse Arm to Centre (Normal)	0.317″	
27	Reverse Arm to Centre (Maximum)	0.350″	

Dockstader Valve Gear Programme			
Data for Inside Admission Walschaert's Valve Gear Simulation (3.5″ Gauge "Britannia")			
Data for Loco Frame Dimensions and Linkage		**Britannia**	**Optimised**
Data for Locomotive Cylinders			
28	Cylinder C/L to Driving Wheel C/L	11.687	
29	Cylinder C/L above Driving Wheel C/L	n/a	
30	Cylinder Bore Diameter	1.250″	
31	Cylinder Bore Length	2.281″	
32	Cylinder Steam Port Width	0.125″	
33	Piston C/L to Crosshead C/L	3.692″	
34	Piston Width	0.438″	
35	Valve Stem C/L to Cylinder C/L	1.375″	
36	Valve Bore (Diameter)	0.688″	
37	Valve Bore Length	3.625″	
38	Valve Bore Centre to Valve Port	0.625″	
39	Valve Bore Port Width Rear	0.156″	
40	Valve Bore Port Width Front	0.156″	
41	Valve centre to inside edge of inside ring	0.500″	0.515″
42	Valve End Width (Outside edges of rings) Rear	0.266″	
43	Valve End Width (Outside edges of rings) Front	0.266″	
44	Valve Centre to Joint (Valve Stem)	2.873″	

Valve events from 3.5″ Gauge Inside Admission Walschaert's Valve Gear – LBSC's BR Class 7 Britannia 4-6-2 (Original)														
Notch	Angle	Travel	Compression		Admission		Cut-Off		Exhaust		Port Opening		Lead	
			Front	Back	Front	Back	Front	Back	Front	Back	Front	Back	Front	Back
0.8	3.506″	0.543″	6.5%	5.2%	0.0%	0.1%	78.4%	77.2%	93.2%	91.6%	0.130″	0.163″	-0.001″	0.001″
0.7	3.461″	0.487″	8.4%	6.8%	0.0%	0.1%	73.7%	71.7%	91.4%	89.5%	0.107″	0.130″	-0.001″	0.001″
0.6	3.423″	0.441″	8.4%	6.8%	0.0%	0.1%	68.9%	65.8%	89.4%	87.5%	0.087″	0.104″	-0.001″	0.001″
0.5	3.385″	0.394″	10.5%	10.6%	0.0%	0.1%	60.6%	58.2%	85.0%	81.9%	0.066″	0.078″	-0.001″	0.001″
0.4	3.346″	0.349″	15.4%	12.7%	0.0%	0.1%	49.8%	48.3%	82.5%	79.1%	0.045″	0.054″	-0.001″	0.001″
0.3	3.306″	0.308″	20.9%	17.5%	0.0%	0.1%	35.5%	36.7%	74.1%	72.9%	0.026″	0.032″	-0.001″	0.001″
0.2	3.266″	0.276″	27.1%	22.9%	0.0%	0.1%	17.5%	19.8%	64.7%	63.0%	0.011″	0.015″	-0.001″	0.001″
0.1	3.226″	0.256″	37.0%	35.3%	0.0%	0.1%	4.4%	5.2%	54.5%	52.5%	0.002″	0.004″	-0.001″	0.001″
0	3.185″	0.250″	44.0%	45.5%	___	0.4%	___	0.9%	44.0%	45.5%	-0.001″	0.001″	-0.001″	0.001″
-0.1	3.144″	0.259″	33.6%	32.1%	0.0%	0.1%	6.3%	9.9%	54.5%	56.0%	0.003″	0.006″	-0.001″	0.001″
-0.2	3.103″	0.282″	23.9%	22.9%	0.0%	0.1%	20.3%	25.3%	67.9%	66.4%	0.013″	0.019″	-0.001″	0.001″
-0.3	3.062″	0.316″	18.1%	15.0%	0.0%	0.1%	39.0%	41.2%	77.1%	76.1%	0.029″	0.037″	-0.001″	0.001″
-0.4	3.022″	0.358″	12.8%	10.6%	0.0%	0.1%	56.0%	54.1%	82.5%	81.9%	0.049″	0.059″	-0.001″	0.001″
-0.5	2.982″	0.409″	10.5%	8.6%	0.0%	0.1%	67.9%	62.9%	87.3%	84.6%	0.076″	0.083″	-0.001″	0.001″
-0.6	2.942″	0.470″	8.4%	6.8%	0.0%	0.1%	76.2%	69.5%	89.4%	87.2%	0.111″	0.109″	-0.001″	0.001″
-0.7	2.903″	0.539″	6.5%	5.2%	0.0%	0.1%	81.7%	74.3%	91.4%	89.5%	0.156″	0.133″	-0.001″	0.001″
-0.8	2.865″	0.634″	6.5%	3.9%	0.0%	0.1%	85.7%	78.1%	93.2%	91.6%	0.225″	0.159″	-0.001″	0.001″
Width of Port			0.156″	Length of Lap			0.125″	Exhaust Lap			0.016″	Dockstader Simulation Inside Admission Walschaert's Valve Gear		
Length of Eccentric Rod			3.769″	Length of Return Crank			0.977″	Expansion Link Angles			-27.04° +16.52°			
Compression Score		60/68	Admission Score		0/68	Cut–Off Score		43/68	Exhaust Score		64/68	Overall Score 167/272		
Percentage		88.2%	Percentage		0%	Percentage		63.2%	Percentage		94.1%	Percentage 61.4%		

Valve events from 3.5″ Gauge Inside Admission Walschaert's Valve Gear – LBSC's BR Class 7 Britannia 4-6-2 (Optimised)														
Notch	**Angle**	**Travel**	**Compression**		**Admission**		**Cut-Off**		**Exhaust**		**Port Opening**		**Lead**	
			Front	Back	Front	Back	Front	Back	Front	Back	Front	Back	Front	Back
0.8	3.506″	0.530″	8.4%	6.8%	0.1%	0.1%	79.6%	79.1%	93.2%	91.6%	0.136″	0.174″	0.014″	0.016″
0.7	3.461″	0.478″	10.5%	8.6%	0.5%	0.4%	75.3%	74.1%	91.4%	89.5%	0.116″	0.142″	0.014″	0.016″
0.6	3.423″	0.432″	12.8%	10.6%	0.5%	0.4%	70.3%	68.8%	89.4%	87.2%	0.096″	0.116″	0.014″	0.016″
0.5	3.385″	0.388″	15.4%	12.7%	0.5%	0.4%	63.5%	62.0%	87.3%	84.6%	0.077″	0.091″	0.014″	0.016″
0.4	3.346″	0.344″	18.1%	15.0%	0.5%	0.4%	53.9%	53.8%	85.0%	81.9%	0.057″	0.067″	0.014″	0.016″
0.3	3.306″	0.305″	23.9%	22.9%	1.2%	1.0%	42.0%	41.9%	77.1%	76.1%	0.039″	0.046″	0.014″	0.016″
0.2	3.266″	0.274″	30.3%	28.9%	2.1%	1.7%	26.2%	28.2%	71.1%	69.7%	0.025″	0.029″	0.014″	0.016″
0.1	3.226″	0.256″	40.5%	42.1%	3.3%	2.7%	14.1%	14.5%	57.9%	59.5%	0.017″	0.019″	0.014″	0.016″
0	3.185″	0.250″	51.0%	52.5%	6.5%	6.8%	7.5%	7.8%	47.5%	49.0%	0.014″	0.016″	0.014″	0.016″
-0.1	3.144″	0.259″	37.0%	38.7%	2.1%	2.7%	15.1%	17.2%	61.3%	63.0%	0.018″	0.021″	0.014″	0.016″
-0.2	3.103″	0.280″	27.1%	25.9%	1.2%	1.0%	29.3%	32.0%	74.1%	72.9%	0.027″	0.033″	0.014″	0.016″
-0.3	3.062″	0.314″	20.9%	20.2%	1.2%	1.0%	46.2%	47.0%	79.8%	79.1%	0.043″	0.050″	0.014″	0.016″
-0.4	3.022″	0.355″	18.1%	15.0%	0.5%	0.4%	60.9%	57.4%	85.0%	81.9%	0.064″	0.071″	0.014″	0.016″
-0.5	2.982″	0.404″	12.8%	10.6%	0.5%	0.4%	72.0%	65.4%	89.4%	87.2%	0.090″	0.094″	0.014″	0.016″
-0.6	2.942″	0.465″	10.5%	8.6%	0.5%	0.4%	78.6%	71.5%	91.4%	89.5%	0.126″	0.119″	0.014″	0.016″
-0.7	2.903″	0.537″	8.5%	6.8%	0.1%	0.4%	83.6%	76.8%	93.2%	91.6%	0.174″	0.143″	0.014″	0.016″
-0.8	2.865″	0.642″	6.5%	5.2%	0.1%	0.1%	87.1%	79.8%	94.8%	93.5%	0.255″	0.167″	0.014″	0.016″
Width of Port			0.156″	Length of Lap			0.110″	Exhaust Lap			0.000″	Dockstader Simulation Inside Admission Walschaert's Valve Gear		
Length of Eccentric Rod			3.769″	Length of Return Crank			0.977″	Expansion Link Angles			-27.04° +16.52°			
Compression Score		63/68	Admission Score		68/68	Cut-Off Score		53/68	Exhaust Score		63/68	Overall Score 247/272		
Percentage		92.6%	Percentage		100%	Percentage		77.9%	Percentage		92.6%	Percentage 90.8%		

5in Gauge

2.5 *Simplex*

Dockstader Data input for Outside Admission Walschaert's Valve Gear 5″ Gauge. Martin Evans' 0-6-0 Tank "Simplex" and modifications by the late Don Ashton		
Data for Loco Frame Dimensions and Linkage	**Original**	**Don Ashton**
Driving Wheel Diameter	4.375″	4.375″
Stroke of Crank	2.125″	2.125″
Length of Connecting Rod (Main Rod)	7.000″	7.000″
Drop Link to Crosshead C/L (Vertical)	0.813″	**0.976″**
Drop Link to Crosshead C/L (Backset)	0.000″	0.000″
Union Link	1.250″	**1.175**
Combination Lever (Upper Portion)	0.312″	**0.313″**
Combination Lever (Upper Portion Backset)	0.000″	**0.000″**
Combination Lever (Lower Portion)	2.375″	**2.563″**
Radius Rod	4.000″	**3.875″**
Link Radius	4.000″	**3.875″**
Link Crank (Vertical)	1.473″	**1.461″**
Link Crank (Backset)	0.285″	**0.500″**
Eccentric Rod	3.938″	**3.717″**
Eccentric Crank Circle Diameter	1.076″	**0.936″**
Eccentric Crank Length	1.188″	**1.158″**
Radius Rod Extension Length	0.938″	0.938″
Reversing Arm (Vertical)	1.375″	1.375″
Reversing Arm (Backset)	0.000″	0.000″
Lifting Arm (Length)	1.312″	**1.438″**
Lifting Link (Length)	0.000″	0.000″
Reverse Arm Pivot (Vertical)	1.500″	1.500″
Reverse Arm Pivot (Horizontal)	2.250″	2.250
Link Centre Pivot (Vertical)	1.468″	1.468″
Link Centre Pivot (Horizontal)	4.188″	4.188″
Reverse Arm to Centre (Normal)	0.750″	0.750″
Reverse Arm to Centre (Maximum)	0.813″	0.813″

Dockstader Data input for Outside Admission Walschaert's Valve Gear 5″ Gauge. Martin Evans' 0-6-0 Tank "Simplex" and modifications by the late Don Ashton		
Data for Loco Frame Dimensions and Linkage	**Original**	**Don Ashton**
Data for Locomotive Cylinders		
Cylinder C/L to Driving Wheel C/L	11.312″	11.312″
Cylinder C/L above Driving Wheel C/L	0.000″	0.000″
Cylinder Bore Diameter	1.500″	1.500″
Cylinder Bore Length	3.000″	3.000″
Cylinder Steam Port Width	0.141″	0.141″
Piston C/L to Crosshead C/L	4.312″	4.312″
Piston Width	0.625″	0.625″
Valve Surface to Cylinder C/L (Horizontal)	1.250″	1.250″
Valve Stem C/L to Cylinder C/L	1.750″	1.750″
Exhaust Port Width	0.438″	0.438″
Bridge Width (Rear)	0.156″	0.156″
Bridge Width (Front)	0.156″	0.156″
Steam Port Width (Rear)	0.156″	0.156″
Steam Port Width (Front)	0.156″	0.156″
Valve Chamber Height	0.875″	0.875″
Valve Chamber Length	2.063″	2.063″
Valve Height	0.813″	0.813″
Valve Length (Rear Half)	0.656″	**0.650″**
Valve Length (Front Half)	0.656″	**0.650″**
Valve Exhaust Cavity (Rear Half)	0.375″	0.375″
Valve Exhaust Cavity (Front Half)	0.375″	0.375″
Valve Centre to Joint (Valve Stem)	3.131″	**3.262″**

Valve events from Full Size Outside Admission Walschaert's Valve Gear fitted to 5" Gauge 0-6-0 Tank "Simplex" Original														
Notch	Angle	Travel	Compression		Admission		Cut-Off		Exhaust		Port Opening		Lead	
			Front	Back	Front	Back	Front	Back	Front	Back	Front	Back	Front	Back
0.8	3.000	0.574"	8.7%	8.2%	0.1%	0.1%	77.2%	77.6%	91.8%	91.3%	0.150"	0.174"	0.016"	0.014"
0.7	2.910	0.523"	10.9%	8.2%	0.6%	0.1%	73.0%	73.7%	91.8%	89.1%	0.126"	0.147"	0.016"	0.014"
0.6	2.821	0.473"	10.9%	10.1%	0.6%	0.4%	67.3%	69.6%	89.9%	89.1%	0.103"	0.120"	0.016"	0.014"
0.5	2.731	0.425"	15.9%	14.4%	0.6%	0.4%	60.3%	63.0%	85.6%	84.1%	0.081"	0.094"	0.015"	0.014"
0.4	2.642	0.380"	18.6%	16.9%	0.6%	0.4%	50.4%	53.8%	83.1%	81.4%	0.060"	0.070"	0.015"	0.014"
0.3	2.552	0.339"	24.7%	22.1%	1.3%	0.9%	39.5%	42.7%	77.9%	75.3%	0.042"	0.047"	0.015"	0.014"
0.2	2.463	0.307"	31.2%	31.1%	1.3%	0.9%	26.3%	29.4%	68.9%	68.8%	0.027"	0.030"	0.015"	0.014"
0.1	2.373	0.286"	38.0%	41.0%	2.2%	2.6%	14.6%	14.2%	59.0%	62.0%	0.018"	0.018"	0.014"	0.014"
0	2.284	0.278"	52.1%	51.5%	5.0%	5.0%	6.1%	6.0%	48.5%	47.9%	0.014"	0.014"	0.014"	0.014"
-0.1	2.194	0.281"	41.5%	41.0%	3.5%	2.6%	12.4%	11.5%	59.0%	58.5%	0.016"	0.017"	0.014"	0.014"
-0.2	2.105	0.301"	34.5%	31.1%	1.3%	0.9%	23.7%	21.8%	68.9%	65.5%	0.025"	0.026"	0.013"	0.014"
-0.3	2.015	0.329"	24.7%	22.1%	0.6%	0.9%	39.4%	35.8%	77.9%	75.3%	0.039"	0.040"	0.013"	0.014"
-0.4	1.926	0.364"	18.6%	16.9%	0.6%	0.4%	51.8%	46.9%	83.1%	81.4%	0.057"	0.057"	0.013"	0.014"
-0.5	1.836	0.407"	15.9%	12.2%	0.6%	0.4%	63.4%	56.8%	87.8%	84.1%	0.079"	0.078"	0.013"	0.014"
-0.6	1.747	0.452"	13.3%	10.1%	0.1%	0.4%	70.3%	64.5%	89.9%	86.7%	0.103"	0.099"	0.012"	0.014"
-0.7	1.657	0.502"	10.9%	8.2%	0.1%	0.4%	76.0%	70.4%	91.8%	89.1%	0.129"	0.123"	0.012"	0.014"
-0.8	1.568	0.551"	8.7%	6.5%	0.1%	0.1%	80.1%	75.0%	93.5%	91.3%	0.155"	0.146"	0.012"	0.014"

Post Width	0.156"	Lap of Valve	0.125"	Exhaust Clearance	0.000"	S.P.Bowditch 16th November 2022.
Maximum Link Swing +	20.56°	Minimum Link Swing -	-21.59°	Eccentric Crank Circle Dia.	1.056"	
Length of Eccentric Rod	3.937"	Length of Return Crank	1.188"	Simulated using the late Charlie Dockstader's Walscheart's Simulation		

Compression Score	58 / 68	Admission Score	68 / 68	Cut-Off Score	47 / 68	Exhaust Score	57 / 68	Total Score	230 / 272
Compression %	85.3%	Admission %	100%	Cut-Off %	69.1%	Exhaust %	83.8%	Total %	84.6%

Valve events from Full Size Outside Admission Walschaert's Valve Gear fitted to 5″ Gauge 0-6-0 Tank "Simplex" Don Ashton														
Notch	Angle	Travel	Compression		Admission		Cut-Off		Exhaust		Port Opening		Lead	
			Front	Back	Front	Back	Front	Back	Front	Back	Front	Back	Front	Back
0.8	3.000	0.557″	6.7%	6.5%	0.1%	0.1%	79.1%	79.1%	93.5%	93.3%	0.155″	0.164″	0.010″	0.010″
0.7	2.910	0.507″	8.7%	8.2%	0.1%	0.1%	75.7%	75.1%	91.8%	91.3%	0.130″	0.139″	0.010″	0.010″
0.6	2.821	0.458″	10.9%	10.1%	0.1%	0.1%	70.0%	70.1%	89.9%	89.1%	0.106″	0.114″	0.010″	0.010″
0.5	2.731	0.410″	13.3%	12.2%	0.1%	0.4%	63.4%	63.5%	87.8%	86.1%	0.083″	0.089″	0.010″	0.010″
0.4	2.642	0.363″	15.9%	16.9%	0.6%	0.4%	53.5%	54.7%	83.1%	84.1%	0.060″	0.065″	0.010″	0.010″
0.3	2.552	0.323″	21.6%	22.1%	0.6%	0.4%	41.1%	43.0%	77.9%	78.4%	0.041″	0.044″	0.010″	0.010″
0.2	2.463	0.289″	27.9%	28.0%	0.6%	0.9%	26.9%	29.5%	72.0%	72.1%	0.025″	0.026″	0.010″	0.010″
0.1	2.373	0.267″	38.0%	37.6%	1.3%	1.6%	13.1%	13.9%	62.4%	62.0%	0.014″	0.015″	0.010″	0.010″
0	2.284	0.258″	48.5%	51.5%	5.0%	3.7%	4.8%	4.8%	48.5%	51.5%	0.010″	0.010″	0.010″	0.010″
-0.1	2.194	0.264″	41.5%	41.0%	2.2%	1.6%	10.7%	10.1%	59.0%	58.5%	0.013″	0.013″	0.010″	0.010″
-0.2	2.105	0.282″	31.2%	31.1%	1.3%	0.9%	23.5%	23.6%	68.9%	68.8%	0.022″	0.022″	0.010″	0.010″
-0.3	2.015	0.312″	24.7%	22.1%	0.6%	0.4%	39.5%	37.1%	77.8%	75.3%	0.037″	0.037″	0.010″	0.010″
-0.4	1.926	0.349″	18.6%	16.9%	0.6%	0.4%	51.8%	50.8%	83.1%	81.4%	0.056″	0.055″	0.010″	0.010″
-0.5	1.836	0.392″	13.3%	12.2%	0.1%	0.4%	63.4%	60.1%	87.8%	86.7%	0.077″	0.077″	0.010″	0.010″
-0.6	1.747	0.439″	10.9%	10.1%	0.1%	0.1%	70.4%	67.5%	89.9%	89.1%	0.101″	0.100″	0.010″	0.010″
-0.7	1.657	0.486″	8.7%	8.2%	0.1%	0.1%	76.1%	73.2%	91.8%	91.3%	0.125″	0.123″	0.010″	0.010″
-0.8	1.568	0.535″	8.7%	6.5%	0.1%	0.1%	80.2%	77.4%	93.5%	91.3%	0.150″	0.147″	0.010″	0.010″

Port Width	0.156″	Lap of Valve	0.119″	Exhaust Clearance	0.000″	S.P.Bowditch 17th November 2022.
Maximum Link Swing +	19.01°	Minimum Link Swing -	-18.75°	Eccentric Crank Circle Dia.	0.936″	
Length of Eccentric Rod	3.717″	Length of Return Crank	1.158″	Simulated using the late Charlie Dockstader's Walscheart's Simulation		

Compression Score	65 / 68	Admission Score	68 / 68	Cut-Off Score	62 / 68	Exhaust Score	66 / 68	Total Score	261/272
Compression %	95.6%	Admission %	100%	Cut-Off %	91.2%	Exhaust %	97.1%	Total %	95.9%

2.6 *Speedy*

Input Data for Prof B. Hall's Walschaert's Valve Gear Simulation Software			
Locomotive: 5″ Gauge BR Western Region 15XX Class Pannier Tank Bill Hall's "Speedy"			
No.	**Dimension**	**LBSC's Original**	**Bill Hall's "Speedy"**
0	Length of the Connecting Rod	7.500″	7.500″
1	Stroke (Radius)	1.063″	1.063″
2	Radius of the Return Crank	0.475″	**0.600″**
3	Lifting Arm Length	1.094″	1.094″
4	Radius Rod Length	4.250″	4.250″
5	Overall Length of Combination Lever	2.750″	2.750″
6	Top section of Combination Lever	0.375″	0.375″
7	Drop Link Length	0.750″	0.750″
8	Anchor Link Length	1.063″	1.063″
9	Steam Lap	0.156″	**0.125″**
10	Exhaust Lap	0.000″	**-0.005″**
11	Expansion Link Pivot to Eccentric Rod Connection	1.601″	1.601″
12	Backset Angle of Expansion Link Connection	5.599″	5.599°
13	Horizontal Distance from Crosshead Pin to Drop Link	0.000″	0.000″
14	Vertical Distance Piston Rod to Valve Spindle	1.500″	1.500″
15	Horizontal Distance from Expansion Link Pivot to Axle Centre	4.250″	4.250″
16	Vertical Distance from Expansion Link Pivot to Axle Centre	1.813″	**1.875″**
17	Horizontal Distance from Driving Axle to Weighshaft	2.625″	2.625″
18	Vertical Distance from Driving Axle to Weighshaft	1.813″	**1.875″**
19	Port Width	0.188″	**0.150″**
22	Length of Suspension Link	n/a	n/a
23	Suspension point on radius rod to combination lever joint	n/a	n/a
20	Valve Type – Outside Admission – Inside Admission	Inside	Inside
21	Radius Rod – Slot and Die – Suspension Link	Slot and Die	Slot & Die
	Calculated length of Eccentric Rod	4.139″	**4.136″**
	Calculated length of Return Crank	1.187″	**1.256″**
	Anti-clockwise swing of Expansion Link	-18.05°	**-23.2°**
	Clockwise swing of Expansion Link	16.87°	**21.67°**

Valve events Inside Admission Walschaert's Valve Gear fitted to 5″ Gauge W.R. 15XX 0-6-0 PT (LBSC "Speedy" Original)														
Notch	Angle	Travel	Compression		Admission		Cut-Off		Exhaust		Port Opening		Lead	
			Front	Back	Front	Back	Front	Back	Front	Back	Front	Back	Front	Back
0.8	40.0°	0.429″	14.88%	12.19%	0.165%	0.109%	51.34%	50.49%	87.81%	85.12%	0.051″	0.066″	-0.011″	-0.011″
0.7	35.0°	0.407″	16.63%	13.94%	0.210%	0.141%	45.14%	45.66%	86.06%	83.37%	0.041″	0.054″	-0.011″	-0.011″
0.6	30.0°	0.383″	18.89%	16.24%	0.276%	0.187%	37.46%	39.61%	83.76%	81.11%	0.030″	0.041″	-0.011″	-0.011″
0.5	25.0°	0.361″	21.78%	19.31%	0.405%	0.265%	28.35%	32.06%	80.69%	78.22%	0.020″	0.029″	-0.011″	-0.011″
0.4	20.0°	0.338″	25.45%	23.35%	0.691%	0.434%	25.45%	23.35%	76.65%	74.55%	0.010″	0.016″	-0.011″	-0.011″
0.3	15.0°	0.318″	30.06%	28.58%	1.922%	0.962%	7.12%	12.03%	71.42%	69.94%	0.001″	0.005″	-0.011″	-0.011″
0.2	10.0°	0.303″	37.71%	35.09%					64.91%	64.29%	-0.005″	-0.004″	-0.011″	-0.011″
0.1	5.0°	0.294″	42.42%	42.64%					57.36%	57.58%	-0.009″	-0.009″	-0.011″	-0.011″
0	0.0°	0.290″	49.35%	50.05%					49.95%	50.65%	-0.011″	-0.011″	-0.011″	-0.011″
-0.1	-5.0°	0.293″	41.63%	42.22%					57.78%	58.37%	-0.010″	-0.009″	-0.011″	-0.011″
-0.2	-10.0°	0.302″	34.77%	34.78%					65.22%	65.23%	-0.006″	-0.004″	-0.011″	-0.011″
-0.3	-15.0°	0.318″	29.02%	28.35%	2.213%	1.051″	6.34%	12.42%	71.65%	70.98%	0.001″	0.005″	-0.011″	-0.011″
-0.4	-20.0°	0.337″	24.36%	23.16%	0.754%	0.483%	17.56%	23.94%	76.84%	75.64%	0.009″	0.016″	-0.011″	-0.011″
-0.5	-25.0°	0.360″	20.68%	19.14%	0.443%	0.299%	27.71%	33.78%	80.86%	79.32%	0.019″	0.029″	-0.011″	-0.011″
-0.6	-30.0°	0.383″	17.79%	16.10%	0.299%	0.211%	36.89%	41.78%	83.90%	82.21%	0.029″	0.042″	-0.011″	-0.011″
-0.7	-35.0°	0.407″	15.55%	13.81%	0.233%	0.164%	44.71%	48.16%	86.19%	84.45%	0.040″	0.055″	-0.011″	-0.011″
-0.8	-40.0°	0.429″	13.81%	12.08%	0.186%	0.132%	51.05%	53.19%	87.92%	86.19%	0.050″	0.067″	-0.011″	-0.011″
Port Width			0.188″	Lap of Valve			0.156″	Exhaust Clearance			0.000″	Prof. Bill Hall Simulator.		
Compression Score		62/68	Admission Score		48/48	Cut-Off Score		29/48	Exhaust Score		63/68	Total Score 202/232		
Compression %		91.2%	Admission %		100%	Cut-Off %		60.0%	Exhaust %		92.6%	Overall 87.0%		

Valve events Inside Admission Walschaert's Valve Gear fitted to 5″ Gauge W.R. 15XX 0-6-0 PT (Prof. Bill Hall's "Speedy")														
Notch	Angle	Travel	Compression		Admission		Cut-Off		Exhaust		Port Opening		Lead	
			Front	Back	Front	Back	Front	Back	Front	Back	Front	Back	Front	Back
0.8	40.0°	0.509″	10.38%	7.819%	0.299%	0.207%	73.6%	68.74%	91.14%	88.33%	0.121″	0.138″	0.020″	0.020″
0.7	35.0°	0.475″	11.86%	9.118%	0.362%	0.247%	69.51%	65.06%	89.68%	86.70%	0.104″	0.121″	0.020″	0.020″
0.6	30.0°	0.439″	13.84%	10.92%	0.454%	0.316%	64.01%	60.32%	87.68%	84.53%	0.087″	0.102″	0.020″	0.020″
0.5	25.0°	0.403″	16.48%	13.44%	0.588%	0.413%	56.59%	54.22%	84.88%	81.66%	0.070″	0.083″	0.020″	0.020″
0.4	20.0°	0.368″	20.02%	17.01%	0.819%	0.591%	46.88%	46.41%	80.94%	77.83%	0.054″	0.064″	0.020″	0.020″
0.3	15.0°	0.337″	24.75%	22.11%	1.227%	0.912%	35.17%	36.62%	75.40%	72.76%	0.040″	0.047%	0.020″	0.020″
0.2	10.0°	0.302″	30.98%	29.20%	2.020%	1.583%	23.18%	25.2%	67.81%	66.17%	0.029″	0.032″	0.020″	0.020″
0.1	5.0°	0.295″	38.92%	38.41%	3.644%	3.130%	13.34%	14.20%	58.22%	57.86%	0.022″	0.023″	0.020″	0.020″
0	0.0°	0.290″	47.66%	48.11%	6.933%	6.702%	6.990%	6.851%	48.44%	48.90%	0.020″	0.020″	0.020″	0.020″
-0.1	-5.0°	0.295″	37.79%	37.96%	3.542%	3.060%	13.14%	14.58%	58.65%	59.03%	0.022″	0.023″	0.020″	0.020″
-0.2	-10.0°	0.310″	29.65%	28.93%	1.934%	1.545%	22.43%	26.12%	68.05%	67.57%	0.028″	0.032″	0.020″	0.020″
-0.3	-15.0°	0.335″	23.41%	21.94%	1.162%	0.888%	33.86%	38.31%	75.55%	74.22%	0.039″	0.046″	0.020″	0.020″
-0.4	-20.0°	0.365″	18.75%	16.92%	0.768%	0.574%	45.45%	48.67%	81.03%	79.22%	0.051″	0.064″	0.020″	0.020″
-0.5	-25.0°	0.399″	15.33%	13.40%	0.542%	0.400%	55.41%	56.76%	84.91%	82.94%	0.066″	0.083″	0.020″	0.020″
-0.6	-30.0°	0.433″	12.81%	10.93%	0.415%	0.304%	63.09%	62.94%	87.66%	85.69%	0.081″	0.102″	0.019″	0.019″
-0.7	-35.0°	0.468″	10.95%	9.165%	0.324%	0.236%	68.80%	67.4%	89.63%	87.73%	0.097″	0.121″	0.019″	0.019″
-0.8	-40.0°	0.501″	9.561%	7.895%	0.272%	0.199%	72.98%	71.23%	91.05%	89.26%	0.112″	0.139″	0.019″	0.019″
Port Width			0.150″	Lap of Valve			0.125″	Exhaust Clearance			-0.005″	Prof. Bill Hall Simulator.		
Compression Score		62/68	Admission Score		68/68		Cut-Off Score		56/68	Exhaust Score		62/68	Total Score 248/272	
Compression %		91.2%	Admission %		100%		Cut-Off %		82.4%	Exhaust %		91.2%	Overall 91.2%	

2.7 *Didcot*

Stephenson's Valve Gear Simulation by the late Prof. Bill Hall			
Locomotive: 5″ Gauge GWR 48XX / B.R. 14XX 0-4-2 Tank Locomotive "Didcot" by late Neville Evans			
Key	**Description**	**Didcot Original**	**Don Ashton**
1	Connecting Rod (Length)	6.750″	6.750″
2	Main Crank (Radius)	1.000″	1.000″
3	Eccentric (Radius)	0.375″	0.375″
4	Angle of Advance (Degrees)	7.000°	**9.500°**
5	Eccentric Rod (Length)	4.250″	**4.160″**
6	Eccentric Rod Pin to Expansion Link Centre Line (Length)	0.000″	0.000″
7	Crank Axle Centre to Weighshaft Centre (Horizontal Length)	3.110″	**2.349″**
8	Crank Axle Centre to Weighshaft Centre (Vertical Distance)	1.375″	**-1.340″**
9	Lifting Arm (Radius)	1.250″	**2.000″**
10	Lifting Link (Length)	1.375″	**2.269″**
11	Distance Between Pins on Expansion Link (Length)	1.500″	**1.720″**
12	Circumference Displacement of Lifting Pin on Expansion Link	0.000″	0.660″
13	Radial Displacement of Lifting Pin on Expansion Link	0.000″	0.000″
14	Lap of Steam Valve	0.094″	**0.120″**
15	Exhaust Lap of Steam Valve	0.000″	0.000″
16	Port Width	0.125″	0.125″
17	Rocker Arm Ratio	n/a	n/a
Additional Information			
	Valve Type Outside OR Inside Admission	Outside	Outside
	Rocker (No Rocker) – (Type A Rocker) – (Type B Rocker)	No Rocker	No Rocker

Valve events from Outside Admission Stephenson's Valve Gear fitted to 5" Gauge 0-4-2 Tank Loco (Neville Evans' Original Design of "Didcot")														
Notch	Angle	Travel	Compression		Admission		Cut-Off		Exhaust		Port Opening		Lead	
			Front	Back	Front	Back	Front	Back	Front	Back	Front	Back	Front	Back
0.8	15.000°	0.398"	7.06%	6.424%	0.049%	0.034%	77.35%	78.94%	93.58%	92.94%	0.085"	0.125"	0.006"	0.006"
0.7	13.125°	0.368"	8.96%	8.27%	0.125%	0.087%	71.58%	74.97%	91.72%	91.20%	0.069"	0.111"	0.009"	0.009"
0.6	11.250°	0.334"	10.97%	10.79%	0.263%	0.184%	63.67%	70.10%	89.21%	89.03%	0.054"	0.092"	0.012"	0.012"
0.5	9.375°	0.303"	13.70%	14.24%	0.489%	0.347%	52.73%	64.05%	85.76%	86.30%	0.042"	0.073"	0.014"	0.014"
0.4	7.500°	0.277"	17.15%	19.06%	0.859%	0.630%	38.83%	56.40%	80.94%	82.85%	0.033"	0.056"	0.016"	0.016"
0.3	5.625°	0.255"	21.52%	25.58%	1.489%	1.164%	25.22%	46.83%	74.15%	78.46%	0.026"	0.041"	0.017"	0.017"
0.2	3.750°	0.238"	27.10%	35.12%	2.545%	2.215%	15.21%	34.41%	64.88%	72.90%	0.021"	0.029"	0.018"	0.018"
0.1	1.875°	0.229"	34.18%	46.43%	4.327%	4.465%	8.877%	21.17%	53.57%	65.82%	0.019"	0.022"	0.018"	0.019"
0	0.000°	0.226"	43.01%	57.73%	7.266%	9.408%	7.266%	5.112%	42.27%	56.99%	0.019"	0.019"	0.019"	0.019"
-0.1	-1.875°	0.229"	32.79%	46.66%	2.946%	5.192%	11.89%	19.07%	53.34%	67.21%	0.020"	0.021"	0.018"	0.019"
-0.2	-3.750°	0.239"	25.49%	36.10%	1.707%	2.55%	18.76%	32.81%	63.90%	74.51%	0.024"	0.027"	0.017"	0.018"
-0.3	-5.625°	0.257"	19.95%	27.02%	0.997%	1.327%	28.29%	46.06%	72.98%	80.05%	0.030"	0.039"	0.016"	0.017"
-0.4	-7.500°	0.278"	15.75%	20.09%	0.569%	0.720%	40.12%	55.64%	79.91%	84.25%	0.038"	0.052"	0.015"	0.016"
-0.5	-9.375°	0.306"	12.52%	15.05%	0.314%	0.393%	52.29%	64.53%	84.95%	87.48%	0.048"	0.070"	0.013"	0.014"
-0.6	-11.250°	0.336"	10.00%	11.41%	0.163%	0.209%	62.58%	70.72%	88.59%	90.00%	0.059"	0.089"	0.010"	0.012"
-0.7	-13.125°	0.378"	8.029%	8.755%	0.077%	0.101%	70.46%	75.58%	91.24%	91.97%	0.072"	0.108"	0.008"	0.010"
-0.8	-15.000°	0.400"	6.462%	6.782%	0.026%	0.044%	76.37%	79.49%	93.22%	93.54%	0.087"	0.125"	0.004"	0.007"

Port Width Front Port Width Rear	0.125" 0.125"	Lap of Valve Front Lap of Valve Rear	0.094" 0.094"	Exhaust Clearance Front Rear	0.000" 0.000"	Outside Admission Stephenson's Simulation Prof.Bill Hall 11/01/2023
Link Centre Pivot Back Set	**0.000"**	Angle of Advance of Eccentric	**7.000°**	Length of Lifting Arm	**1.250"**	

Compression Score	45/68	Admission Score	66/68	Cut-Off Score	26/68	Exhaust Score	45/68	Total Points	**182/272**
Percentage	66.2%	Percentage	97.1%	Percentage	38.2%	Percentage	66.2%	Percentage	**66.91%**

Valve events from Outside Admission Stephenson's Valve Gear fitted to 5″ Gauge 0-4-2 Tank Loco (Don Ashton's Modified Design of "Didcot")														
Notch	Angle	Travel	Compression		Admission		Cut-Off		Exhaust		Port Opening		Lead	
			Front	Back	Front	Back	Front	Back	Front	Back	Front	Back	Front	Back
0.8	+18.0°	0.481″	5.242%	4.173%	-0.226%	-0.144%	80.18%	77.53%	95.83%	94.76%	0.116″	0.125″	-0.018″	-0.017″
0.7	+15.75°	0.449″	7.166%	5.898%	-0.107%	-0.066%	74.30%	71.99%	94.10%	92.83%	0.097″	0.112″	-0.011″	-0.010″
0.6	+13.50°	0.410″	9.704%	8.319%	-0.028%	-0.017%	66.65%	65.07%	91.68%	90.03%	0.079″	0.091″	-0.004″	-0.003″
0.5	+11.25°	0.375″	13.03%	11.77%	0.001%	0.002%	57.04%	56.52%	88.23%	86.92%	0.063″	0.072″	0.001″	0.002″
0.4	+9.00°	0.342″	17.58%	16.70%	0.073%	0.071%	45.77%	46.21%	83.30%	82.42%	0.048″	0.054″	0.005″	0.006″
0.3	+6.75°	0.314″	23.53%	23.60%	0.280%	0.263%	33.87%	34.48%	76.40%	76.47%	0.036″	0.038″	0.009″	0.010″
0.2	+4.50°	0.292″	31.27%	32.65	0.775%	0.762%	22.71%	22.51%	67.35%	68.73%	0.026″	0.026″	0.013″	0.013″
0.1	+2.25°	0.277″	40.85%	43.12%	1.886%	2.246%	13.54%	12.19%	56.88%	59.15%	0.019″	0.018″	0.015″	0.014″
0	0.0°	0.271″	51.67%	53.70%	4.408%	5.747%	7.11%	5.29%	51.67%	53.70%	0.016″	0.015″	0.016″	0.015″
-0.1	-2.25°	0.274″	36.69%	37.56%	3.344%	1.984%	9.88%	17.64%	62.44%	63.31%	0.016″	0.018″	0.015″	0.014″
-0.2	-4.50°	0.290″	28.52%	28.14%	1.459%	0.717%	19.95%	32.21%	71.86%	71.48%	0.020″	0.030″	0.014″	0.012″
-0.3	-6.75°	0.316″	21.86%	20.62%	0.579%	0.235%	34.00%	45.36%	79.38%	79.14%	0.029″	0.047″	0.012″	0.010″
-0.4	-9.00°	0.349″	16.56%	14.94%	0.189%	0.059%	48.41%	56.09%	85.06%	83.44%	0.041″	0.068″	0.009″	0.006″
-0.5	-11.25°	0.387″	12.40%	10.75%	0.031%	0.005%	60.40%	64.74%	89.25%	87.60%	0.056″	0.091″	0.004″	0.001″
-0.6	-13.50°	0.430″	9.18%	7.70%	-0.009%	-0.002%	69.60%	71.73%	92.30%	90.82%	0.074″	0.116″	-0.001″	-0.004″
-0.7	-15.75°	0.458″	6.69%	5.48%	-0.065%	-0.082%	76.57%	77.39%	94.52%	93.31%	0.093″	0.125″	-0.008″	-0.011″
-0.8	-18.0°	0.479″	4.79%	3.86%	-0.176%	-0.174%	81.87%	82.00%	96.15%	95.21%	0.114″	0.125″	-0.016″	-0.020″

Port Width Front Port Width Rear	0.125″ 0.125″	Lap of Valve Front Lap of Valve Rear	0.120″ 0.120″	Exhaust Clearance Front Rear	0.000″ 0.000″	Outside Admission Stephenson's Simulation Prof.Bill Hall 10/01/2023
Link Centre Pivot Back Set	**0.000″**	Angle of Advance of Eccentric	**13.00°**	Length of Lifting Arm	**2.000″**	

Compression Score	66/68	Admission Score	68/68	Cut-Off Score	51/68	Exhaust Score	66/68	Total Score	251/272
Percentage	97.1%	Percentage	100%	Percentage	75.0%	Percentage	97.1%	Percentage	**92.3%**

2.8 *Class 2 78xxx*

Dockstader Valve Gear Programme			
Data for Inside Admission Walschaert's Valve Gear (5″ Gauge B.R. Class 2MT 2-6-0)			
Data for Loco Frame Dimensions and Linkage		**Original**	**Optimised**
1	Driving Wheel Diameter	5.250″	5.250″
2	Stroke of Crank	2.125″	2.125″
3	Length of Connecting Rod (Main Rod)	7.625″	7.625″
4	Drop Link to Crosshead C/L (Vertical)	1.000″	**0.938″**
5	Drop Link to Crosshead C/L (Back-set)	0.000″	0.000″
6	Union Link	1.000″	**1.063″**
7	Combination Lever (Upper Portion)	0.563″	**0.375″**
8	Combination Lever (Upper Portion Back-set)	0.000″	0.000″
9	Combination Lever (Lower Portion)	2.813″	**2.750″**
10	Radius Rod	3.6875″	**3.609″**
11	Link Radius	3.6875″	**3.609″**
12	Link Crank (Vertical)	1.403″	**1.500″**
13	Link Crank (Backset)	0.195″	**0.250″**
14	Eccentric Rod	4.973″	**4.780″**
15	Eccentric Crank Circ	1.125″	1.125″
16	Eccentric Crank Length	1.266″	**1.238″**
17	Radius Rod Extension Length	1.000″	1.000″
18	Reversing Arm (Vertical)	1.375″	1.375″
19	Reversing Arm (Back-set)	-0.253″	**-0.369″**
20	Lifting Arm (Length)	1.625″	1.625″
21	Lifting Link (Length)	0.000″	0.000″
22	Reverse Arm Pivot (Vertical)	1.672″	1.672″
23	Reverse Arm Pivot (Horizontal)	2.716″	2.716″
24	Link Centre Pivot (Vertical)	1.912″	**1.839″**
25	Link Centre Pivot (Horizontal)	4.978″	**4.875″**
26	Reverse Arm to Centre (Normal)	0.500″	**0.625″**
27	Reverse Arm to Centre (Maximum)	0.563″	**0.688″**

Dockstader Valve Gear Programme			
Data for Inside Admission Walschaert's Valve Gear (5″ Gauge B.R. Class 2MT 2-6-0)			
Data for Loco Frame Dimensions and Linkage		Original	Optimised
Data for Locomotive Cylinders			
28	Cylinder C/L to Driving Wheel C/L	12.219″	12.219″
29	Cylinder C/L above Driving Wheel C/L	0.000″	0.000″
30	Cylinder Bore Diameter	1.438″	1.438″
31	Cylinder Bore Length	2.906″	2.906″
32	Cylinder Steam Port Width	0.156″	0.156″
33	Piston C/L to Crosshead C/L	4.594″	4.594″
34	Piston Width	0.563″	0.563″
35	Valve Stem C/L to Cylinder C/L	1.750″	1.750″
36	Valve Bore (Diameter)	0.750″	0.750″
37	Valve Bore Length	4.250″	4.250″
38	Valve Bore Centre to Valve Port	0.594″	0.594″
39	Valve Bore Port Width Rear	0.156″	0.156″
40	Valve Bore Port Width Front	0.156″	0.156″
41	Valve centre to inside edge of inside ring	0.438″	**0.481″**
42	Valve End Width (Outside edges of rings) Rear	0.312″	**0.269″**
43	Valve End Width (Outside edges of rings) Front	0.312″	**0.269″**
44	Valve Centre to Joint (Valve Stem)	3.576″	**3.719″**

Valve events Inside Admission Walschaert's Valve Gear fitted to 5″ Gauge B.R. Standard Class 2 2-6-0 by Don Young (Orig.)														
Notch	Angle	Travel	Compression		Admission		Cut-Off		Exhaust		Port Opening		Lead	
			Front	Back	Front	Back	Front	Back	Front	Back	Front	Back	Front	Back
0.8	2.997″	0.636″	13.1%	10.3%	0.6%	0.4%	73.8%	69.9%	89.7%	89.2%	0.148″	0.176″	0.021″	0.023″
0.7	2.919″	0.587″	13.1%	10.3%	0.6%	0.4%	69.4%	66.4%	89.7%	86.9%	0.124″	0.151″	0.021″	0.023″
0.6	2.841″	0.537″	15.7%	12.3%	0.6%	0.4%	63.4%	61.0%	87.7%	84.3%	0.100″	0.125″	0.021″	0.023″
0.5	2.763″	0.489″	18.5%	17.0%	0.6%	0.4%	56.2%	54.9%	83.0%	81.5%	0.078″	0.099″	0.021″	0.023″
0.4	2.681″	0.444″	21.4%	19.6%	1.2%	0.9%	44.6%	46.5%	80.4%	78.6%	0.058″	0.074″	0.021″	0.023″
0.3	2.606″	0.409″	27.6%	25.3%	1.2%	0.9%	32.8%	36.9%	74.7%	72.4%	0.043″	0.054″	0.021″	0.022″
0.2	2.528″	0.380″	34.3%	34.6%	2.2%	1.7%	20.9%	25.2%	68.6%	65.7%	0.031″	0.037″	0.021″	0.022″
0.1	2.450″	0.362″	41.2%	41.3%	3.4%	3.7%	12.1%	14.0%	58.7%	58.8%	0.024″	0.026″	0.021″	0.022″
0	2.372″	0.355″	51.8%	51.8%	6.6%	6.6%	5.9%	6.2%	48.2%	48.2%	0.021″	0.022″	0.021″	0.022″
-0.1	2.294″	0.360″	41.2%	41.3%	3.4%	3.7%	11.8%	13.7%	58.7%	58.8%	0.023″	0.025″	0.021″	0.022″
-0.2	2.216″	0.377″	34.3%	31.4%	2.2%	1.7%	20.2%	25.1%	68.6%	69.1%	0.030″	0.035″	0.021″	0.021″
-0.3	2.138″	0.407″	27.6%	25.3%	1.2%	0.9%	32.7%	39.1%	74.7%	75.6%	0.042″	0.053″	0.021″	0.021″
-0.4	2.060″	0.445″	21.4%	19.6%	1.2%	0.9%	46.3%	49.8%	80.4%	78.6%	0.059″	0.074″	0.021″	0.021″
-0.5	1.981″	0.497″	15.7%	14.6%	0.6%	0.4%	57.9%	59.6%	85.4%	84.3%	0.081″	0.104″	0.021″	0.021″
-0.6	1.903″	0.553″	13.1%	12.3%	0.6%	0.4%	66.5%	66.4%	87.7%	86.9%	0.106″	0.135″	0.021″	0.021″
-0.7	1.825″	0.614″	10.8%	10.3%	0.6%	0.4%	72.8%	71.2%	89.7%	89.2%	0.134″	0.168″	0.021″	0.020″
-0.8	1.747″	0.677″	10.8%	8.3%	0.6%	0.4%	77.5%	75.4%	91.7%	91.4%	0.164″	0.201″	0.021″	0.020″
Port Width			0.156″	Lap of Valve			0.156″	Exhaust Clearance			-0.001			

Compression Score	61/68	Admission Score	68/68	Cut-Off Score	50/68	Exhaust Score	64/68	Total Score 243/272
Compression %	89.7%	Admission %	100%	Cut-Off %	73.5%	Exhaust %	94.1%	Overall 89.3%

Valve events Inside Admission Walschaert's Valve Gear fitted to 5″ Gauge B.R. Standard Class 2 2-6-0 by Don Young (Mk5x)														
Notch	Angle	Travel	Compression		Admission		Cut-Off		Exhaust		Port Opening		Lead	
			Front	Back	Front	Back	Front	Back	Front	Back	Front	Back	Front	Back
0.8	3.604″	0.528″	8.6%	6.6%	0.1%	0.1%	80.3%	76.3%	93.4%	91.4%	0.142″	0.160″	0.014″	0.015″
0.7	3.525″	0.477″	10.8%	8.3%	0.6%	0.4%	75.8%	71.8%	91.7%	89.2%	0.117″	0.134″	0.014″	0.015″
0.6	3.447″	0.430″	13.1%	10.3%	0.6%	0.4%	69.9%	66.8%	89.7%	86.9%	0.094″	0.110″	0.014″	0.015″
0.5	3.369″	0.384″	15.7%	12.3%	0.6%	0.4%	62.0%	60.0%	87.7%	84.3%	0.072″	0.086″	0.014″	0.015″
0.4	3.291″	0.343″	18.5%	17.0%	0.6%	0.4%	51.4%	51.1%	83.0%	81.5%	0.053″	0.064″	0.014″	0.015″
0.3	3.213″	0.308″	24.4%	22.4%	1.2%	0.9%	39.3%	40.2%	77.6%	75.6%	0.037″	0.045″	0.014″	0.015″
0.2	3.135″	0.279″	30.9%	31.4%	1.2%	1.2%	24.3%	27.5%	68.6%	69.1%	0.024″	0.029″	0.014″	0.015″
0.1	3.057″	0.262″	41.2%	41.3%	3.4%	2.6%	13.0%	16.0%	58.7%	58.8%	0.017″	0.019″	0.014″	0.015″
0	2.979″	0.254″	51.8%	51.8%	6.6%	6.6%	5.8%	6.1%	48.2%	48.2%	0.013″	0.015″	0.013″	0.015″
-0.1	2.900″	0.259″	41.2%	41.3%	3.4%	3.7%	12.0%	13.8%	58.7%	58.8%	0.015″	0.018″	0.013″	0.015″
-0.2	2.822″	0.273″	30.9%	31.4%	2.2%	1.7%	20.7%	26.7%	68.6%	69.1%	0.021″	0.026″	0.013″	0.015″
-0.3	2.744″	0.297″	24.4%	25.3%	1.2%	0.9%	33.1%	39.3%	74.7%	75.6%	0.031″	0.040″	0.013″	0.015″
-0.4	2.666″	0.325″	21.4%	19.6%	0.6%	0.9%	45.3%	49.7%	80.4%	78.6%	0.043″	0.056″	0.013″	0.015″
-0.5	2.588″	0.359″	15.7%	14.6%	0.6%	0.4%	56.7%	57.7%	85.4%	84.3%	0.058″	0.075″	0.013″	0.015″
-0.6	2.510″	0.396″	13.1%	12.3%	0.6%	0.4%	64.4%	64.4%	87.7%	86.9%	0.075″	0.095″	0.013″	0.015″
-0.7	2.432″	0.433″	10.8%	10.3%	0.6%	0.4%	70.7%	70.0%	89.7%	89.2%	0.092″	0.115″	0.013″	0.015″
-0.8	2.354″	0.472″	10.8%	8.3%	0.1%	0.4%	75.6%	73.8%	91.7%	89.2%	0.110″	0.136″	0.013″	0.015″
Port Width			0.156″	Lap of Valve			0.113″	Exhaust Clearance			0.000	S.P.B 5th August 2023.		
Compression Score		64/68	Admission Score		68/68	Cut-Off Score		54/68	Exhaust Score		64/68	Total Score 250/272		
Compression %		94.1%	Admission %		100%	Cut-Off %		79.4%	Exhaust %		94.1%	Overall 91.9%		

2.9 *Manor*

Stephenson's Valve Gear Simulation by the late Prof. Bill Hall			
Locomotive: 5″ Gauge GWR 78XX 4-6-0 Tender Locomotive "Torquay Manor" by the late Martin Evans			
Key	**Description**	**Original by M. Evans**	**Modified by D. Ashton**
1	Connecting Rod (Length)	11.313″	11.313″
2	Main Crank (Radius)	1.313″	1.313″
3	Eccentric (Radius)	0.330″	0.330″
4	Angle of Advance (Degrees)	28.0 deg.	**18.50 deg.**
5	Eccentric Rod (Length)	4.230″	4.230″
6	Eccentric Rod Pin to Expansion Link Centre Line (Length)	0.395″	0.395″
7	Crank Axle Centre to Weighshaft Centre (Horizontal Length)	3.313″	3.313″
8	Crank Axle Centre to Weighshaft Centre (Vertical Distance)	1.875″	1.875″
9	Lifting Arm (Radius)	1.313″	**1.140″**
10	Lifting Link (Length)	1.875″	1.875″
11	Distance Between Pins on Expansion Link (Length)	1.250″	1.250″
12	Circumference Displacement of Lifting Pin on Expansion Link	0.000″	0.000″
13	Radial Displacement of Lifting Pin on Expansion Link	0.032″	**0.155″**
14	Lap of Steam Valve	0.156″	**0.140″**
15	Exhaust Lap of Steam Valve	0.000″	0.000″
16	Port Width	0.156″	0.156″
17	Rocker Arm Ratio	1.027	1.027
	Valve Type Outside OR Inside Admission	Outside	Outside
	Rocker (No Rocker) – (Type A Rocker) – (Type B Rocker)	Type A	Type A

Prof. Bill Hall's Stephenson Valve Gear Simulation														
Valve events from Outside Admission Stephenson's Valve Gear fitted to 5" Gauge GWR "Manor" Class 4-6-0 Tender Locomotive by Martin Evans - ORIGINAL														
Notch	Angle	Travel	Compression		Admission		Cut-Off		Exhaust		Port Opening		Lead	
			Front	Back	Front	Back	Front	Back	Front	Back	Front	Back	Front	Back
0.8	24 degs	0.600"	12.16%	6.94%	0.265%	0.140%	78.07%	64.87%	93.06%	87.84%	0.148"	0.140"	0.021"	0.021"
0.7	21 degs	0.559"	15.12%	8.60%	0.533%	0.281%	73.78%	58.70%	91.40%	84.88%	0.127"	0.120"	0.026"	0.027"
0.6	18 degs	0.522"	18.78%	10.69%	0.974%	0.509%	68.38%	51.50%	89.31%	81.22%	0.108"	0.102"	0.033"	0.032"
0.5	15 degs	0.489"	23.90%	13.34%	1.649%	0.859%	61.55%	43.31%	86.66%	76.71%	0.091"	0.086"	0.037"	0.036"
0.4	12 degs	0.458"	28.78%	16.70%	2.705%	1.396%	52.99%	34.51%	83.30%	71.22%	0.074"	0.072"	0.041"	0.040"
0.3	9 degs	0.435"	35.33%	20.95%	4.344%	2.220%	42.77%	25.85%	79.05%	64.07%	0.062"	0.061"	0.044"	0.044"
0.2	6 degs	0.419"	42.89%	26.31%	6.871%	3.490%	31.87%	18.26%	73.69%	57.11%	0.053"	0.054"	0.046"	0.046"
0.1	3 degs	0.409"	51.10%	32.91%	10.72%	5.428%	22.02%	12.32%	67.09%	48.90%	0.048"	0.049"	0.047"	0.047"
0	0 degs	0.407"	59.38%	40.69%	16.34%	8.343%	14.44%	8.058%	59.31%	40.62%	0.048"	0.047"	0.047"	0.048"
-0.1	-3 degs	0.412"	50.81%	32.95%	9.229%	5.168%	24.01%	12.58%	67.05%	49.19%	0.050"	0.050"	0.047"	0.047"
-0.2	-6 degs	0.423"	42.34%	26.35%	5.837%	3.275%	33.41%	18.42%	73.65%	57.66%	0.056"	0.055"	0.045"	0.045"
-0.3	-9 degs	0.441"	34.64%	20.94%	3.675%	2.053%	43.56%	25.87%	79.08%	65.36%	0.066"	0.063"	0.043"	0.043"
-0.4	-12 degs	0.464"	28.07%	16.57%	2.293%	1.263%	53.19%	34.44%	83.43%	71.93%	0.078"	0.074"	0.040"	0.040"
-0.5	-15 degs	0.494"	22.65%	13.13%	1.406%	0.755%	61.51%	43.25%	86.67%	77.35%	0.093"	0.089"	0.036"	0.035"
-0.6	-18 degs	0.528"	18.27%	10.40%	0.826%	0.423%	68.35%	51.47%	89.60%	81.73%	0.111"	0.105"	0.031"	0.030"
-0.7	-21 degs	0.566"	14.74%	8.245%	0.460%	0.218%	73.88%	58.71%	91.75%	85.26%	0.130"	0.124"	0.026"	0.024"
-0.8	-24 degs	0.606"	11.91%	6.531%	0.229%	0.093%	78.35%	64.87%	93.47%	88.09%	0.150"	0.144"	0.020"	0.017"
Port Width			0.156"	Lap of Valve			0.156"	Exhaust Clearance			0.000"			

Compression Score	17/68	Admission Score	57/68	Cut-Off Score	17/68	Exhaust Score	17/68	Symmetry Score	108/272
Compression Percentage	25%	Admission Percentage	83.8%	Cut-Off Percentage	25%	Exhaust Percentage	25%	Symmetry %	39.7%

Prof. Bill Hall's Stephenson Valve Gear Simulation														
Valve events from Outside Admission Stephenson's Valve Gear fitted to 5" Gauge GWR "Manor "Class 4-6-0 Tender Locomotive by Don Ashton														
Notch	Angle	Travel	Compression		Admission		Cut-Off		Exhaust		Port Opening		Lead	
			Front	Back	Front	Back	Front	Back	Front	Back	Front	Back	Front	Back
0.8	24 degs	0.551"	7.165%	5.446%	-0.038%	-0.024%	79.13%	75.16%	94.55%	92.83%	0.124"	0.147"	-0.007"	-0.007"
0.7	21 degs	0.504"	9.162%	7.012%	-0.012%	-0.008%	74.24%	70.14%	92.99%	90.84%	0.102"	0.122"	-0.002"	-0.002"
0.6	18 degs	0.459"	11.73%	9.103%	0.013%	0.001%	67.67%	63.88%	90.90%	88.27%	0.081"	0.098"	0.002"	0.002"
0.5	15 degs	0.416"	15.05%	11.93%	0.063%	0.042%	58.74%	55.98%	88.07%	84.95%	0.061"	0.075"	0.006"	0.006"
0.4	12 degs	0.379"	19.35%	15.79%	0.201%	0.135%	46.78%	45.97%	84.21%	80.65%	0.045"	0.054"	0.009"	0.009"
0.3	9 degs	0.348"	24.91%	21.11%	0.498%	0.345%	32.51%	33.62%	78.89%	75.09%	0.031"	0.037"	0.012"	0.012"
0.2	6 degs	0.327"	32.04%	28.38%	1.138%	0.813%	19.26%	20.33%	71.62%	67.96%	0.022"	0.025%	0.014"	0.014"
0.1	3 degs	0.314"	40.95%	37.83%	2.524%	1.920%	9.935%	9.894%	62.17%	59.05%	0.017"	0.017"	0.015"	0.015"
0	0 degs	0.310"	51.38%	48.73%	5.487%	4.649%	4.662%	4.165%	51.27%	48.62%	0.015"	0.015"	0.015"	0.015"
-0.1	-3 degs	0.316"	40.64%	37.83%	2.107%	1.688%	11.22%	10.57%	62.17%	59.36%	0.018"	0.018"	0.015"	0.015"
-0.2	-6 degs	0.330"	31.61%	28.31%	0.949%	0.700%	20.60%	20.79%	71.69%	68.39%	0.024"	0.026"	0.013"	0.013"
-0.3	-9 degs	0.352"	24.46%	20.91%	0.417%	0.290%	33.28%	33.71%	79.09%	75.54%	0.033"	0.039"	0.011"	0.011"
-0.4	-12 degs	0.384"	18.96%	15.48%	0.161%	0.104%	46.92%	45.97%	84.52%	81.04%	0.047"	0.057"	0.009"	0.009"
-0.5	-15 degs	0.420"	14.75%	11.55%	0.046%	0.026%	58.74%	56.00%	88.45%	85.25%	0.063"	0.077"	0.005"	0.005"
-0.6	-18 degs	0.462"	11.52%	8.695%	0.009%	0.004%	67.83%	63.91%	91.30%	88.48%	0.082"	0.100"	0.001"	0.001"
-0.7	-21 degs	0.508"	9.029%	6.588%	-0.015%	-0.014%	74.61%	70.16%	93.41%	90.71%	0.103"	0.125"	-0.003"	-0.004"
-0.8	-24 degs	0.555"	7.090%	5.013%	-0.044%	-0.039%	79.70%	75.17%	94.99%	92.91%	0.125"	0.150"	-0.008"	-0.009"
Port Width			0.156"	Lap of Valve			0.140"	Exhaust Clearance			0.000"			
Compression Score		48/68	Admission Score		68/68	Cut-Off Score		55/68	Exhaust Score		48/68	Symmetry Score		219/272
Compression Percentage		70.6%	Admission Percentage		100%	Cut-Off Percentage		80.9%	Exhaust Percentage		70.6%	Symmetry %		80.5%

7.25in Gauge

2.10 *Dart*

Stephenson's Valve Gear Simulation by the late Prof. Bill Hall			
Locomotive: 7.25″ Gauge GWR 48XX / B.R. 14XX 0-4-2 Tank Locomotive "Dart" by the late Martin Evans			
Key	**Description**	**Dart Original**	**Dart Three**
1	Connecting Rod (Length)	9.000″	9.000″
2	Main Crank (Radius)	1.500″	1.500″
3	Eccentric (Radius)	0.315″	0.315″
4	Angle of Advance (Degrees)	23.0°	**20.0°**
5	Eccentric Rod (Length)	5.563″	**6.000″**
6	Eccentric Rod Pin to Expansion Link Centre Line (Length)	0.438″	**0.000″**
7	Crank Axle Centre to Weighshaft Centre (Horizontal Length)	3.215″	3.215″
8	Crank Axle Centre to Weighshaft Centre (Vertical Distance)	2.000″	2.000″
9	Lifting Arm (Radius)	2.750″	2.750″
10	Lifting Link (Length)	2.000″	**2.938″**
11	Distance Between Pins on Expansion Link (Length)	1.375″	**1.875″**
12	Circumference Displacement of Lifting Pin on Expansion Link	0.000″	**0.938″**
13	Radial Displacement of Lifting Pin on Expansion Link	0.219″	**0.000″**
14	Lap of Steam Valve	0.125″	0.125″
15	Exhaust Lap of Steam Valve	0.000″	0.000″
16	Port Width	0.187″	0.187″
17	Rocker Arm Ratio	n/a	n/a
Additional Information			
	Valve Type Outside OR Inside Admission	Outside	Outside
	Rocker (No Rocker) – (Type A Rocker) – (Type B Rocker)	No Rocker	No Rocker

Valve events from Outside Admission Stephenson's Valve Gear fitted to GWR 0-4-2 Tank 'Dart' Original Version														
Notch	Angle	Travel	Compression		Admission		Cut-Off		Exhaust		Port Opening		Lead	
			Front	Back	Front	Back	Front	Back	Front	Back	Front	Back	Front	Back
0.8	15°	0.625″	4.3%	3.1%	-0.008%	-0.004%	87.6%	84.05%	96.82%	95.6%	0.1875″	0.1875″	-0.002″	-0.001″
0.7	13.1	0.594″	6.07%	4.49%	0.02%	0.018%	83.82%	79.75%	95.51%	93.93%	0.163″	0.181″	0.005″	0.006″
0.6	11.2	0.533″	8.33%	6.32%	0.109%	0.083%	78.65%	74.21%	93.68%	91.66%	0.134″	0.149″	0.013″	0.013″
0.5	9.3	0.475″	11.4%	8.92%	0.306%	0.228%	71.54%	67.00%	91.07%	88.58%	0.107″	0.118″	0.020″	0.019″
0.4	7.5	0.426″	15.4%	12.4%	0.673%	0.513%	62.40%	58.11%	87.53%	84.60%	0.085″	0.091″	0.024″	0.024″
0.3	5.6	0.381″	21.6%	17.86%	1.379%	1.086%	49.88%	46.04%	82.14%	78.84%	0.065″	0.066″	0.028″	0.028″
0.2	3.7	0.347″	29.1%	25.6%	2.721%	2.269%	35.55%	31.54%	74.40%	70.90%	0.049″	0.048″	0.032″	0.031″
0.1	1.8	0.325″	39.6%	36.0%	5.351%	4.844%	22.30%	17.71%	64.00%	60.32%	0.039″	0.036″	0.033″	0.033″
0	0	0.317″	51.8%	47.4%	10.150%	9.979%	12.98%	8.872%	52.52%	48.16%	0.034″	0.033″	0.034″	0.033″
-0.1	-1.8	0.324″	41.2%	35.7%	7.110%	4.250%	18.73%	19.18%	64.28%	59.8%	0.036″	0.038″	0.034″	0.033″
-0.2	-3.7	0.345″	30.9%	24.8%	3.670%	1.989%	32.64%	32.44%	75.18%	69.05%	0.044″	0.051″	0.033″	0.031″
-0.3	-5.6	0.378″	22.9%	17.6%	1.874%	0.948%	48.8%	45.9%	82.94%	77.10%	0.058″	0.070″	0.030″	0.028″
-0.4	-7.5	0.423″	16.8%	11.8%	0.927%	0.442%	62.63%	57.34%	88.19%	83.17%	0.078″	0.095″	0.026″	0.024″
-0.5	-9.3	0.473″	12.5%	8.4%	0.448%	0.197%	72.23%	66.03%	91.57%	87.47%	0.101″	0.122″	0.022″	0.019″
-0.6	-11.2	0.532″	9.15%	5.95%	0.171%	0.067%	79.41%	73.29%	94.05%	90.84%	0.129″	0.153″	0.015″	0.012″
-0.7	-13.1	0.594″	6.65%	4.21%	0.038%	0.014%	84.49%	78.96%	95.78%	93.34%	0.158″	0.186″	0.008″	0.005″
-0.8	-15°	0.624″	4.80%	2.98%	0.001%	-0.008%	88.19%	83.42%	97.02%	95.20%	0.187″	0.187″	0.000″	-0.003″

Width of Steam Port	0.188″	Length of Valve Lap	0.125″	Length of Exhaust Lap	0.000″	Valve Gear Simulation using Bill Hall's Stephenson Program Simon Bowditch 23rd March 2015.
Angle of Advance of Eccentric	23°	Equal Leads Set at Lifting Arm	6.0°	Minimum Angle of Link Clockwise at Mid Gear	-22.6°	
Throw of Eccentrics	0.315″			Maximum Angle of Link Anticlockwise at Mid Gear	22.6°	

Compression Score	41/68	Admission Score	67/68	Cut-Off Score	36/68	Exhaust Score	42/68	Overall Score 186/272
Compression %	60.3%	Admission %	98.5%	Cut-Off %	52.9%	Exhaust %	61.8%	Overall % 68.4%

Valve events from Outside Admission Stephenson's Valve Gear fitted to GWR 0-4-2 Tank 'Dart' Modified Version with Locomotive Links by Don Ashton and Simon Bowditch														
Notch	Angle	Travel	Compression		Admission		Cut-Off		Exhaust		Port Opening		Lead	
			Front	Back	Front	Back	Front	Back	Front	Back	Front	Back	Front	Back
0.8	15°	0.525″	7.32%	5.15%	0.010%	0.005%	78.21%	74.72%	94.48%	92.68%	0.127″	0.148″	0.002″	0.002″
0.7	13.1	0.483″	9.60%	7.50%	0.059%	0.038%	72.16%	69.12%	92.49%	90.40%	0.107″	0.126″	0.008″	0.008″
0.6	11.2	0.444″	12.5%	10.2%	0.196%	0.133%	64.45%	62.35%	89.76%	87.45%	0.089″	0.105″	0.014″	0.014″
0.5	9.3	0.406″	16.3%	14.0%	0.460%	0.327%	54.99%	54.29%	85.97%	83.66%	0.071″	0.085″	0.018″	0.018″
0.4	7.5	0.376″	20.9%	18.9%	0.894%	0.662%	44.82%	45.49%	81.05%	79.07%	0.058″	0.068″	0.022″	0.022″
0.3	5.6	0.348″	27.0%	25.8%	1.702%	1.338%	33.73%	35.33%	74.11%	72.98%	0.046″	0.052″	0.025″	0.025″
0.2	3.7	0.327″	34.5%	34.6%	3.134%	2.693%	23.52%	25.1%	65.40%	65.48%	0.037″	0.040″	0.027″	0.027″
0.1	1.8	0.313″	43.2%	44.4%	5.681%	5.573%	15.1%	16.0%	55.58%	56.71%	0.031″	0.032″	0.029″	0.029″
0	0	0.308″	52.2%	53.7%	9.827%	11.17%	9.42%	9.43%	46.24%	47.71%	0.029″	0.029″	0.029″	0.029″
-0.1	-1.8	0.313″	37.6%	38.7%	5.598%	5.170%	16.38%	20.54%	61.21%	62.36%	0.031″	0.032″	0.029″	0.029″
-0.2	-3.7	0.328″	29.8%	30.2%	3.138%	2.659%	26.08%	32.68%	69.74%	70.19%	0.036″	0.042″	0.028″	0.028″
-0.3	-5.6	0.352″	23.3%	23.1%	1.712%	1.363%	37.59%	44.29%	76.88%	76.66%	0.045″	0.057″	0.025″	0.025″
-0.4	-7.5	0.384″	18.1%	17.4%	0.896%	0.686%	49.07%	54.28%	82.54%	81.87%	0.057″	0.077″	0.022″	0.022″
-0.5	-9.3	0.419″	14.1%	13.3%	0.456%	0.341%	58.67%	62.17%	86.69%	85.19%	0.071″	0.098″	0.019″	0.019″
-0.6	-11.2	0.461″	10.8%	9.9%	0.188%	0.142%	67.02%	69.06%	90.03%	89.11%	0.089″	0.122″	0.014″	0.014″
-0.7	-13.1	0.505″	8.3%	7.4%	0.052%	0.042%	73.76%	74.73%	92.54%	91.70%	0.108″	0.147″	0.008″	0.008″
-0.8	-15°	0.551″	6.2%	5.5%	0.006%	0.007%	79.05%	79.38%	94.43%	93.72%	0.128″	0.173″	0.001″	0.002″

Width of Steam Port	0.188″	Length of Valve Lap	0.125″	Length of Exhaust Lap	0.000″	Valve Gear Simulation using Bill Hall's Stephenson Program
Angle of Advance of Eccentric	20°	Equal Leads Set at Lifting Arm	6.0°	Minimum Angle of Link Clockwise at Mid Gear	-17.63°	
Throw of Eccentrics	0.315″			Maximum Angle of Link Anticlockwise at Mid Gear	17.06°	Simon Bowditch 25th March 2015.

Compression Score	64/68	Admission Score	68/68	Cut-Off Score	51/68	Exhaust Score	65/68	Overall Score 248/272
Compression %	94.1%	Admission %	100%	Cut-Off %	75%	Exhaust %	95.6%	Overall % 91.2%

2.11 *Class 2T 84XXX*

Dockstader Valve Gear Programme			
Data for Inside Admission Walschaert's Valve Gear (7.25″ Gauge BR Class 2 84XXX Tank by D. Stockings)			
Data for Loco Frame Dimensions and Linkage		Original (B)	Modified (E)
1	Driving Wheel Diameter	6.813″	6.813″
2	Stroke of Crank	3.000″	3.000″
3	Length of Connecting Rod (Main Rod)	11.438″	11.438″
4	Drop Link to Crosshead C/L (Vertical)	1.438″	1.438″
5	Drop Link to Crosshead C/L (Back-set)	0.000″	0.000″
6	Union Link	1.625″	1.625″
7	Combination Lever (Upper Portion)	0.590″	**0.558″**
8	Combination Lever (Upper Portion Back-set)	0.000″	0.000″
9	Combination Lever (Lower Portion)	3.835″	3.835″
10	Radius Rod	6.172″	6.172″
11	Link Radius	6.172″	6.172″
12	Link Crank (Vertical)	2.057″	**2.000″**
13	Link Crank (Backset)	0.422″	**0.375″**
14	Eccentric Rod	6.490″	**6.501″**
15	Eccentric Crank Circ	1.810″	**1.624″**
16	Eccentric Crank Length	1.844″	**1.756″**
17	Radius Rod Extension Length	2.125″	2.125″
18	Reversing Arm (Vertical)	1.500″	1.500″
19	Reversing Arm (Back-set)	-0.284″	**-0.210″**
20	Lifting Arm (Length)	2.000″	2.000″
21	Lifting Link (Length)	0.000″	0.000″
22	Reverse Arm Pivot (Vertical)	2.450″	**2.292″**
23	Reverse Arm Pivot (Horizontal)	3.809″	3.809″
24	Link Centre Pivot (Vertical)	2.482″	2.482″
25	Link Centre Pivot (Horizontal)	6.695″	6.695″
26	Reverse Arm to Centre (Normal)	0.563″	**0.718″**
27	Reverse Arm to Centre (Maximum)	0.688″	**0.750″**

Dockstader Valve Gear Programme			
Data for Inside Admission Walschaert's Valve Gear (7.25″ Gauge BR Class 2 84XXX Tank by D. Stockings)			
Data for Loco Frame Dimensions and Linkage		Original (B)	Modified (E)
Data for Locomotive Cylinders			
28	Cylinder C/L to Driving Wheel C/L	18.063″	18.063″
29	Cylinder C/L above Driving Wheel C/L	0.000″	0.000″
30	Cylinder Bore Diameter	2.125″	2.125″
31	Cylinder Bore Length	3.938″	3.938″
32	Cylinder Steam Port Width	0.156″	0.156″
33	Piston C/L to Crosshead C/L	6.625″	6.625″
34	Piston Width	0.750″	0.750″
35	Valve Stem C/L to Cylinder C/L	2.313″	2.313″
36	Valve Bore (Diameter)	1.000″	1.000″
37	Valve Bore Length	5.563″	5.563″
38	Valve Bore Centre to Valve Port	1.594″	1.594″
39	Valve Bore Port Width Rear	0.219″	0.219″
40	Valve Bore Port Width Front	0.219″	0.219″
41	Valve centre to inside edge of inside ring	1.413″	**1.424″**
42	Valve End Width (Outside edges of rings) Rear	0.400″	**0.389″**
43	Valve End Width (Outside edges of rings) Front	0.400″	**0.389″**
44	Valve Centre to Joint (Valve Stem)	5.180″	**5.181″**

Valve events 7.25″ Gauge Inside Admission Walschaert's Valve Gear – BR 84XXX Class 2 2-6-2 Tank – Dockstader ORIGINAL														
Notch	Angle	Travel	Compression		Admission		Cut-Off		Exhaust		Port Opening		Lead	
			Front	Back	Front	Back	Front	Back	Front	Back	Front	Back	Front	Back
0.8	4.621″	0.796″	10.7%	6.7%	0.1%	0.1%	78.1%	73.5%	93.3%	89.3%	0.202″	0.232″	0.020″	0.018″
0.7	4.531″	0.724″	10.7%	8.4%	0.6%	0.1%	73.5%	69.2%	91.6%	89.3%	0.167″	0.195″	0.019″	0.018″
0.6	4.441″	0.653″	13.1%	10.3%	0.6%	0.4%	67.5%	63.9%	89.7%	86.9%	0.133″	0.158″	0.019″	0.018″
0.5	4.351″	0.587″	18.4%	14.7%	0.6%	0.4%	59.5%	57.1%	85.3%	81.6%	0.102″	0.123″	0.019″	0.018″
0.4	4.262″	0.526″	21.3%	17.2%	0.6%	0.4%	49.2%	49.3%	82.8%	78.7%	0.074″	0.090″	0.019″	0.018″
0.3	4.172″	0.474″	27.5%	22.5%	1.2%	0.4%	35.7%	37.7%	77.5%	72.5%	0.051″	0.061″	0.019″	0.018″
0.2	4.082″	0.434″	34.1%	31.6%	1.2%	1.0%	22.7%	24.8%	68.4%	65.9%	0.034″	0.038″	0.019″	0.018″
0.1	3.992″	0.409″	41.0%	41.5%	2.2%	1.7%	12.0%	13.4%	58.5%	59.0%	0.023″	0.024″	0.019″	0.018″
0	3.903″	0.399″	51.5%	52.0%	6.6%	5.1%	4.6%	4.6%	48.0%	48.5%	0.019″	0.018″	0.019″	0.018″
-0.1	3.813″	0.406″	41.0%	41.5%	3.4%	2.6%	10.1%	11.5%	58.5%	59.0%	0.022″	0.022″	0.019″	0.018″
-0.2	3.723″	0.426″	34.1%	31.6%	1.2%	1.0%	19.7%	23.6%	68.4%	65.9%	0.030″	0.034″	0.019″	0.018″
-0.3	3.633″	0.461″	27.5%	25.4%	1.2%	1.0%	32.1%	36.1%	74.6%	72.5%	0.045″	0.054″	0.019″	0.018″
-0.4	3.544″	0.506″	21.3%	19.8%	0.6%	0.4%	45.6%	47.2%	80.2%	78.7%	0.064″	0.080″	0.019″	0.018″
-0.5	3.454″	0.558″	18.4%	14.7%	0.6%	0.4%	56.2%	56.4%	85.3%	81.6%	0.087″	0.109″	0.019″	0.018″
-0.6	3.364″	0.616″	13.1%	12.4%	0.6%	0.4%	64.9%	63.3%	87.6%	86.9%	0.113″	0.141″	0.019″	0.018″
-0.7	3.274″	0.677″	13.1%	10.3%	0.6%	0.1%	71.2%	68.8%	89.7%	86.9%	0.141″	0.174″	0.019″	0.018″
-0.8	3.185″	0.740″	10.7%	8.4%	0.1%	0.1%	75.9%	73.1%	91.6%	89.3%	0.171″	0.207″	0.019″	0.018″
Width of Port			0.219″	Lap of Valve			0.181″	Exhaust Clearance			0.000″	Valve Events Calculated by Simon Bowditch Dockstader Simulation		
Length of Eccentric Rod			6.476″	Length of Return Crank			1.792″	Eccentric Crank Circle Diameter			1.773″			

Compression Score	50/68	Admission Score	68/68	Cut–Off Score	54/68	Exhaust Score	50/68	Overall Score 222/272
Percentage	73.5%	Percentage	100%	Percentage	79.4%	Percentage	73.5%	Percentage 81.6%

Valve events 7.25″ Gauge Inside Admission Walschaert's Valve Gear – BR 84XXX Class 2 2-6-2 Tank – Dockstader Version E														
Notch	Angle	Travel	Compression		Admission		Cut-Off		Exhaust		Port Opening		Lead	
			Front	Back	Front	Back	Front	Back	Front	Back	Front	Back	Front	Back
0.8	4.625″	0.746″	10.7%	8.4%	0.1%	0.4%	77.3%	73.5%	91.6%	89.3%	0.187″	0.219″	0.020″	0.022″
0.7	4.535″	0.679″	10.7%	10.3%	0.6%	0.4%	72.6%	69.2%	89.7%	89.3%	0.155″	0.184″	0.020″	0.022″
0.6	4.445″	0.613″	13.1%	12.4%	0.6%	0.4%	66.4%	63.7%	87.6%	86.9%	0.124″	0.149″	0.020″	0.022″
0.5	4.355″	0.551″	18.4%	14.7%	0.6%	0.4%	58.1%	56.7%	85.3%	81.6%	0.095″	0.116″	0.020″	0.022″
0.4	4.266″	0.495″	21.3%	19.8%	0.6%	0.4%	47.3%	48.1%	80.2%	78.7%	0.069″	0.086″	0.020″	0.021″
0.3	4.176″	0.447″	27.5%	25.4%	1.2%	1.0%	35.3%	37.1%	74.6%	72.5%	0.048″	0.059″	0.020″	0.021″
0.2	4.086″	0.417″	34.1%	31.6%	2.2%	1.7%	22.6%	24.8%	68.4%	65.9%	0.033″	0.038″	0.020″	0.021″
0.1	3.996″	0.388″	41.0%	41.5%	3.4%	2.6%	11.9%	13.5%	58.5%	59.0%	0.023″	0.025″	0.020″	0.021″
0	3.907″	0.380″	52.0%	51.5%	6.6%	6.7%	5.8%	5.9%	48.0%	48.5%	0.020″	0.020″	0.020″	0.020″
-0.1	3.817″	0.387″	41.5%	41.0%	3.4%	2.6%	11.9%	13.5%	58.5%	59.0%	0.023″	0.024″	0.020″	0.020″
-0.2	3.727″	0.409″	31.6%	34.1%	2.2%	1.7%	20.9%	24.4%	68.4%	65.9%	0.032″	0.037″	0.020″	0.020″
-0.3	3.637″	0.442″	22.5%	27.5%	1.2%	1.0%	33.4%	37.1%	77.5%	72.5%	0.046″	0.056″	0.020″	0.020″
-0.4	3.548″	0.483″	21.3%	19.8%	0.6%	0.4%	46.1%	48.0%	80.2%	78.2%	0.064″	0.079″	0.020″	0.019″
-0.5	3.458″	0.534″	18.4%	14.7%	0.6%	0.4%	57.0%	56.8%	85.3%	81.6%	0.087″	0.107″	0.020″	0.019″
-0.6	3.368″	0.588″	13.1%	12.4%	0.6%	0.4%	66.0%	63.7%	87.6%	86.9%	0.112″	0.136″	0.020″	0.019″
-0.7	3.278″	0.644″	13.1%	10.3%	0.6%	0.4%	72.2%	69.0%	89.7%	86.9%	0.138″	0.166″	0.020″	0.019″
-0.8	3.189″	0.703″	10.7%	8.4%	0.6%	0.1%	76.2%	73.2%	91.6%	89.3%	0.166″	0.197″	0.020″	0.018″

Width of Port	0.219″	Lap of Valve	0.170″	Exhaust Clearance	0.000″	Valve Events Calculated by Simon Bowditch Dockstader Simulation
Length of Eccentric Rod	6.501″	Length of Return Crank	1.756″	Eccentric Crank Circle Diameter	1.624″	

Compression Score	55/68	Admission Score	68/68	Cut–Off Score	56/68	Exhaust Score	55/68	Overall Score 234/272
Percentage	80.8%	Percentage	100%	Percentage	82.4%	Percentage	80.8%	Percentage 86.0%

2.12 *Black 5*

Dockstader Valve Gear Programme			
Data for Inside Admission Walschaert's Valve Gear (7.25″ Gauge LMS 5MT 4-6-0)			
Data for Loco Frame Dimensions and Linkage		Highlander	D Ashton
1	Driving Wheel Diameter	8.875″	8.875″
2	Stroke of Crank	3.500″	3.500″
3	Length of Connecting Rod (Main Rod)	16.750″	16.750″
4	Drop Link to Crosshead C/L (Vertical)	1.688″	**1.656″**
5	Drop Link to Crosshead C/L (Back-set)	0.000″	0.000″
6	Union Link	1.719″	**1.400″**
7	Combination Lever (Upper Portion)	0.625″	**0.602″**
8	Combination Lever (Upper Portion Back-set)	0.000″	0.000″
9	Combination Lever (Lower Portion)	4.188″	**4.187″**
10	Radius Rod	9.875″	**9.630″**
11	Link Radius	9.875″	**9.630″**
12	Link Crank (Vertical)	2.394″	**2.263″**
13	Link Crank (Backset)	0.461″	**0.234″**
14	Eccentric Rod	8.188″	**8.370″**
15	Eccentric Crank Circ	2.065″	**1.750″**
16	Eccentric Crank Length	2.032″	**2.000″**
17	Radius Rod Extension Length	2.188″	2.188″
18	Reversing Arm (Vertical)	1.625″	1.625″
19	Reversing Arm (Back-set)	-0.381″	-0.381″
20	Lifting Arm (Length)	2.500″	2.500″
21	Lifting Link (Length)	0.000″	0.000″
22	Reverse Arm Pivot (Vertical)	2.675″	**2.660″**
23	Reverse Arm Pivot (Horizontal)	4.988″	**4.991″**
24	Link Centre Pivot (Vertical)	2.768″	**2.733″**
25	Link Centre Pivot (Horizontal)	8.494″	**8.498″**
26	Reverse Arm to Centre (Normal)	0.875″	0.875″
27	Reverse Arm to Centre (Maximum)	0.891″	0.891″

Dockstader Valve Gear Programme			
Data for Inside Admission Walschaert's Valve Gear (7.25″ Gauge LMS 5MT 4-6-0)			
Data for Loco Frame Dimensions and Linkage		**Highlander**	**D Ashton**
Data for Locomotive Cylinders			
28	Cylinder C/L to Driving Wheel C/L	24.000	24.000″
29	Cylinder C/L above Driving Wheel C/L	0.000″	0.000″
30	Cylinder Bore Diameter	2.250″	2.250″
31	Cylinder Bore Length	4.500″	4.500″
32	Cylinder Steam Port Width	0.219″	0.219″
33	Piston C/L to Crosshead C/L	7.250″	7.250″
34	Piston Width	0.875″	0.875″
35	Valve Stem C/L to Cylinder C/L	2.375″	2.375″
36	Valve Bore (Diameter)	1.250″	1.250″
37	Valve Bore Length	7.000″	7.000″
38	Valve Bore Centre to Valve Port	2.031″	2.031″
39	Valve Bore Port Width Rear	0.219″	0.219″
40	Valve Bore Port Width Front	0.219″	0.219″
41	Valve centre to inside edge of inside ring	1.812″	**1.825″**
42	Valve End Width (Outside edges of rings) Rear	0.438″	**0.425″**
43	Valve End Width (Outside edges of rings) Front	0.438″	**0.425″**
44	Valve Centre to Joint (Valve Stem)	5.620″?	**5.872″**

Valve events from 7.25″ Gauge Inside Admission Walschaert's Valve Gear – "Highlander" L.M.S. "5MT" Class 4-6-0 (Original)														
Notch	Angle	Travel	Compression		Admission		Cut-Off		Exhaust		Port Opening		Lead	
			Front	Back	Front	Back	Front	Back	Front	Back	Front	Back	Front	Back
0.8	6.192″	1.166″	8.3%	5.3%	0.5%	0.1%	85.2%	79.0%	94.7%	91.7%	0.366″	0.362″	0.045″	0.014″
0.7	6.083″	1.044″	10.5%	6.9%	0.5%	0.1%	81.4%	75.2%	93.1%	89.5%	0.303″	0.303″	0.040″	0.014″
0.6	5.973″	0.922″	10.5%	8.6%	0.5%	0.1%	76.6%	70.3%	91.4%	89.5%	0.241″	0.243″	0.035″	0.013″
0.5	5.863″	0.806″	15.3%	10.6%	0.5%	0.1%	69.9%	63.8%	89.4%	84.7%	0.182″	0.186″	0.031″	0.012″
0.4	5.755″	0.700″	18.0%	12.8%	0.5%	0.1%	60.3%	55.3%	82.2%	82.0%	0.129″	0.133″	0.026″	0.011″
0.3	5.645″	0.604″	23.8%	17.6%	0.5%	0.1%	46.3%	43.6%	82.4%	76.2%	0.082″	0.084″	0.021″	0.011″
0.2	5.536″	0.528″	30.2%	26.0%	0.5%	0.4%	28.4%	28.2%	74.0%	69.8%	0.045″	0.045″	0.017″	0.010″
0.1	5.426″	0.476″	40.4%	38.8%	1.2%	0.4%	11.6%	11.7%	61.2%	59.6%	0.019″	0.019″	0.012″	0.019″
0	5.317″	0.454″	50.9%	52.6%	2.1%	1.7%	2.0%	2.4%	47.4%	49.1%	0.007″	0.009″	0.007″	0.009″
-0.1	5.207″	0.462″	36.9%	38.8%	0.1%	0.4%	6.3%	9.8%	61.2%	63.1%	0.009″	0.015″	0.003″	0.008″
-0.2	5.098″	0.497″	27.0%	29.0%	0.0%	0.1%	19.3%	27.1%	71.0%	73.0%	0.022″	0.037″	-0.001″	0.007″
-0.3	4.988″	0.555″	20.8%	20.2%	0.0%	0.1%	36.2%	44.1%	79.8%	79.2%	0.045″	0.072″	-0.006″	0.007″
-0.4	4.879″	0.628″	15.3%	15.1%	0.0%	0.1%	52.5%	57.2%	84.9%	84.7%	0.075″	0.115″	-0.010″	0.006″
-0.5	4.770″	0.712″	10.5%	10.6%	0.0%	0.1%	64.2%	67.0%	89.4%	89.5%	0.112″	0.162″	-0.014″	0.005″
-0.6	4.660″	0.803″	8.3%	8.6%	0.0%	0.1%	72.1%	73.5%	91.4%	91.7%	0.153″	0.212″	-0.018″	0.005″
-0.7	4.551″	0.895″	6.4%	6.9%	0.0%	0.1%	77.5%	78.0%	93.1%	93.6%	0.195″	0.262″	-0.022″	0.004″
-0.8	4.442″	0.986″	6.4%	5.3%	0.0%	0.1%	81.3%	81.7%	94.7%	93.6%	0.237″	0.311″	-0.025″	0.004″

Width of Port	0.219″	Lap of Valve	0.219″	Exhaust Clearance	0.000″	Valve Events Calculated by Simon Bowditch 29th January 2020.
Length of Eccentric Rod	8.142″	Length of Return Crank	2.024″			

Compression Score	55/68	Admission Score	68/68	Cut–Off Score	41/68	Exhaust Score	58/68	Overall Score 222/272
Percentage	80.9%	Percentage	100%	Percentage	60.3%	Percentage	85.3%	Percentage 81.6%

Valve events from 7.25″ Gauge Inside Admission Walschaert's Valve Gear "Highlander" L.M.S."5MT" Class 4-6-0 (D Ashton)														
Notch	Angle	Travel	Compression		Admission		Cut-Off		Exhaust		Port Opening		Lead	
			Front	Back	Front	Back	Front	Back	Front	Back	Front	Back	Front	Back
0.8	6.225″	1.036″	6.4%	5.3%	0.1%	0.1%	83.6%	80.3%	94.7%	93.6%	0.301″	0.323″	0.014″	0.014″
0.7	6.116″	0.931″	8.3%	6.9%	0.1%	0.1%	80.2%	76.6%	93.1%	91.7%	0.249″	0.270″	0.014″	0.014″
0.6	6.006″	0.828″	10.5%	8.6%	0.1%	0.1%	74.8%	71.3%	91.4%	89.5%	0.198″	0.218″	0.014″	0.014″
0.5	5.897″	0.731″	12.8%	10.6%	0.1%	0.1%	66.2%	63.7%	89.4%	87.2%	0.150″	0.169″	0.014″	0.014″
0.4	5.788″	0.641″	15.3%	15.1%	0.1%	0.1%	57.7%	55.8%	84.9%	84.7%	0.107″	0.122″	0.014″	0.014″
0.3	5.678″	0.561″	20.8%	20.2%	0.5%	0.4%	43.8%	44.1%	79.8%	79.2%	0.069″	0.080″	0.014″	0.014″
0.2	5.569″	0.499″	27.0%	26.0%	0.5%	0.4%	28.1%	30.1%	74.0%	73.0%	0.041″	0.046″	0.014″	0.014″
0.1	5.459″	0.457″	36.9%	38.8%	1.2%	1.0%	12.5%	14.0%	61.2%	63.1%	0.022″	0.023″	0.014″	0.014″
0	5.350″	0.440″	50.9%	49.1%	3.3%	2.7%	4.3%	4.6%	50.9%	49.1%	0.014″	0.014″	0.014″	0.014″
-0.1	5.241″	0.447″	40.4%	42.2%	2.1%	1.7%	8.2%	9.6%	57.8%	59.6%	0.017″	0.018″	0.014″	0.014″
-0.2	5.131″	0.478″	30.2%	32.2%	0.5%	0.4%	19.7%	22.6%	67.8%	69.8%	0.031″	0.035″	0.014″	0.014″
-0.3	5.022″	0.526″	23.8%	23.0%	0.5%	0.4%	34.8%	38.2%	77.0%	76.2%	0.052″	0.062″	0.014″	0.014″
-0.4	4.913″	0.588″	18.0%	17.6%	0.5%	0.1%	48.7%	51.0%	82.4%	82.0%	0.080″	0.096″	0.014″	0.014″
-0.5	4.803″	0.660″	15.3%	12.8%	0.1%	0.1%	60.1%	60.5%	87.2%	84.7%	0.113″	0.135″	0.014″	0.014″
-0.6	4.694″	0.737″	12.8%	10.6%	0.1%	0.1%	68.5%	67.2%	89.4%	87.2%	0.149″	0.176″	0.014″	0.014″
-0.7	4.584″	0.405″	10.5%	8.6%	0.1%	0.1%	74.5%	72.9%	91.4%	89.5%	0.186″	0.219″	0.014″	0.014″
-0.8	4.475″	0.896″	8.3%	6.9%	0.1%	0.1%	78.4%	76.9%	93.1%	91.7%	0.224″	0.260″	0.014″	0.014″
Width of Port			0.219″	Lap of Valve			0.206″	Exhaust Clearance			0.000″	Valve Events Calculated by Simon Bowditch 19th January 2023.		
Length of Eccentric Rod			8.370″	Length of Return Crank			2.000″	Eccentric Crank Circle Dia.			1.750″			
Compression Score		65/68	Admission Score		68/68	Cut-Off Score		59/68	Exhaust Score		66/68	Total Score 258/272		
Percentage		95.6%	Percentage		100%	Percentage		86.8%	Percentage		97.1%	Percentage 94.9%		

BIBLIOGRAPHY

Amsbury, Roy, 'Small Boiler Fittings and how to make them' *Model Engineer* Jan.1973

Ashton. D.L., *Stephenson's Valve Gear* - Camden Miniature Steam services

Ashton. D.L., *Walschaerts Valve Gear* - Camden Miniature Steam services

Bray, Stan, *Grinding, Honing and Polishing* - Special Interest Model Books (SI)

Brown, D.A.G., *Miniature Injectors Inside and Out* TEE Publishing

Cain, Tubal, *Milling Operations in the Lathe* -Nexus Special Interest

Cain, Tubal, *The Model Engineer's Handbook* -Nexus Special Interest

Evans, Martin, 'Building a Garden Railway' *Model Engineer* July 2018

Evans, Martin, *Manual of Model Steam Locomotive Construction* - Model and Allied Publications (MAP)

Ewins, J., 'An Experimental Model' *EIM* February 1982

Ewins, J., 'Testing a Locomotive Boiler' *Model Engineer* March 1966

Farmer, Alec F., *Model Locomotive Boilermaking* - Alec Farmer, Marsden Green

Greenly, Henry, *Model steam Locomotives* Camden Miniature Steam services

Harris, K.N., *Model Boilers and Boilermaking* - Model and Allied Publications (MAP)

Hewson, Doug, 'Building BR Mark1 Coaches in 5in Gauge' *Model Engineer* March 2011

Keiller, C.M., 'Small Injectors' *Journal of the SMEE* Volume 3 No 5 January1962

Lawrence, D.E., 'Making Small Live Steam Injectors' *Model Engineer* April to July 1975

LBSC, *LBSC's Shop, Shed and Road* - TEE Publishing

LBSC, *The Live Steam Book* - TEE Publishing

Palmer, Basil, 'Miniature Live Steam Injectors' *Model Engineer* May 1976

Parkin, Keith, *Mark 1 Coaching Stock of British Railways* - Historical Model Railway Society

Ride on Railways www.rideonrailways.co.uk

Sparey, L.H., *A Man and his Lathe* – George Newnes Ltd London

Sparey, L.H., *The Amateur's Lathe* - George Newnes Ltd London

The Model Engineers Workshop Manual - TEE Publishing

White, H.E., *Small Locomotives* - TEE Publishing